A Discourse on Boxes

of

the Sixteenth, Seventeenth and Eighteenth Centuries.

by

Anthony J. Conybeare.

Published in 1992 by
ROSCA PUBLICATIONS
Units 7/10, Hanley Workshops
Hanley Swan, Worcestershire

British Library Cataloguing-in-Publication Data:
A catalogue record for this book is available from the British
Library.

ISBN 0 9517678 1 X

Designed and produced by the Author.
Printed and bound in Great Britain by

HARTNOLLS Ltd.

CONTENTS

Dedication.

To Dorothy May and Robert Harcourt
Conybeare

INTRODUCTION.

Had the inventor of The Box been able to take out a patent he would surely have made a fortune. The sheer usefulness of the device is astonishing; where would our present civilisation be without it? At one end of the scale we keep matches in boxes, at the other most of us live in them. Large, medium, or small boxes have served the human race for perhaps ten millenia, and no doubt will continue to do so.

I doubt whether anything is ever invented until a need for it arises, and we can only speculate as to what important event gave rise to the need for the first box, but wood was probably the first malleable raw material to be used by man, and to work with wood it is necessary to have developed tools. In any early culture tools are valued, if not venerated, and I cannot believe that a man with tools would not make a box to keep them in. Once tools could be contained, why not other things?

A boat is a box that floats, and early boats were made from logs of wood arduously hollowed out, either by burning or excavation. Dug-out chests were in use right up to Medieval times, when they might just as easily have been made from planks, but no doubt they were deliberately made heavy as a security measure; it would have been difficult for thieves to run far carrying perhaps a quarter-ton of oak, plus whatever had been locked within.

Making a small dug-out box however, was probably not worth the effort, and the true box had to wait until some genius discovered that wood could be split easily along the grain to make planks. This discovery was probably more important in terms of the advancement of the human race than Neil Armstrong's heavily publicised and hideously expensive step down to the surface of the moon. Cleft planks made it possible to construct seaworthy boats for the first time, and the leap from coracle to cargo vessel at last made long distance oceanic trade possible. Trade involved the carriage of goods and the need to keep things secure and apart; commerce also generated cash and bullion which needed to be locked away.

With the invention of the cleft plank, the box had finally arrived. Arrange six of them together and you have the prototype of all the boxes that have ever been made. The need to join wood to wood undoubtedly led to the development of that other useful device, the nail. Invent the chisel and the saw and you are well on the way to the dove-tail joint (much employed in ancient Egypt). Invent the hinge and the lock, and you have taken the box as far as you can along its evolutionary path.

One thing remains to complete the natural history of the box; the decoration.

The human race has always had an extraordinary urge to embellish; given a perfectly good cave wall even prehistoric man discovered a compulsion to decorate it; my parents' generation when given good plain ceilings and walls in their houses could not wait to encrust them with an extraordinary wallpaper which looked as if it had broken out in boils. There has hardly been any period in the development of human culture when decoration of some sort or other has not been applied to otherwise plain surfaces. There appears to be a deep-rooted restlessness about the human spirit which at all ages has led people to decorate everything in sight, from the humblest of their personal possessions to the largest buildings.

Much decoration in early days was symbolic, using devices which carried a message to those who could understand it. The early cave paintings are thought to have had magical significance; the lavish Church decorations of the Middle Ages carried religious messages to people who could not read. Decoration of houses or furniture conveyed the status of the owner.

The carved decoration of the small ordinary domestic boxes in use by the household during the three centuries covered in this book was no different from that used on larger pieces of furniture, we find the same motifs recurring on table-rails, chair backs, court cupboards and beds. When we look at the houses of the period we find them on overmantles, friezes above panelling, and even outside on bargeboards and gables. In the churches they occur on pews, pulpits and screens.

These little decorated boxes have come down to us in considerable numbers, testifying to the robust nature of their construction and to the care of their subsequent owners. For years no serious collector of English oak furniture paid much heed to "Bible Boxes", even thirty years or so ago they tended to pass forlornly from dealer to dealer.

I believe that their image has suffered in the past hundred years from their association with the Victorian custom of using them to contain the family Bible. This was not their original purpose, and (at the risk of offending many people) may I suggest that this is the reason for the lack of interest in them for so many years. The word "Victorian" conjures up mental images of sombre furnishings, and dark rooms, and a monarch who always wore black and was unamused. Most Bibles are black, and our image today of Victorian middle-class domesticity is hardly colourful. Black is not a colour which most people want in their homes; until quite recently black lacquer longcase clocks, although horologically splendid, were difficult to sell. When one considers the boxes it seems unfair that these delightful little articles of domestic furniture, so useful to their original owners, should become burdened with a leaden image, and one purpose of this book is to attempt to replace this.

Each of these little boxes is unique in that it was a "one-off"; it would have been technically impossible for the carver to exactly duplicate his work even if he had wanted to do so; equally, the box maker was quite content with constant variations in the dimensions of his work. All the work that went into a box was by hand; mechanical sawing in England was still two hundred years away from the earlier boxes.

These little boxes were made at a time when their makers were drawing on experience gained in the working of wood over countless generations; much of the carving exhibits skill equal to that of the great woodcarvers of Church furniture in the preceding centuries. The quality of some of the work is more remarkable when one considers that it was carried out with the most basic of tools, the shape of which had not changed since Roman times.

Most of the boxes which we encounter today have been excellently cared for by their various owners down through the generations; the patina on many boxes is higher than that on other larger examples of English furniture, due perhaps to them having been polished more frequently as they were conveniently to hand.

Most are still affordable as they have not yet been granted the status to which I believe they are entitled. To my knowledge, it has not yet been profitable to fake them.

Many excellent books have been written in the last hundred years about English oak furniture, but many have only a small section covering these delightful little articles of everyday use. The time has surely come when they deserve a book of their own.

A.J.Conybeare.   June 1992

Carved "stopped fluting" on the front panel of a late Elizabethan/early Jacobean table box. Found also on fixed woodwork in churches. Originally however it was a pagan pre-Christian motif of great antiquity.

The last of the Gothic; linenfold decoration on a small joined box of the sixteenth century.

The side panels carry a typically Tudor decoration, but the diamond was to continue well into the next century as a popular motif, particularly on the larger panels in chair-backs and court cupboards.

Scrolled "S" curve strapwork on an early 17th Century table box, one of the "new" motifs.

Box no. 1.

ONE

The Classical origins of some early carved motifs.

It is surprising that after the severe, monumental style of furniture that had developed during the Gothic period in England, when architects and artists sought new, lighter, and more attractive motifs, they turned not to the elegant designs of the Celtic period which was so to speak on their doorstep, but to the emblems of great antiquity which came to them via the Renaissance in Italy.

By the second half of the sixteenth century, designs used in the decoration of buildings and furniture almost all employed motifs imported into this country, and almost all were Classical in origin.

There can be little doubt that this seed had fallen on fertile soil, by the sixteenth century the education of the literate classes had been wholly Classical, with Greek and Latin taught as a matter of course, as preparation for entry into the church or the professions.

Until the invention of the printing press, dissemination of ideas and designs was painfully slow, hand copying being the only method of duplication. However, with the development of engraving in the second half of the fifteenth century, the rapid transmission of images was at last possible. When Italian artists began to create a new fashion from ancient designs the means to distribute them was already at hand.

The rediscovery of ancient Roman decorative motifs which had remained buried in the ruins of houses for over a thousand years was to have an impact on European culture which persists until the present day, but the effect on the designers of the Elizabethan and Jacobean period was virtually immediate. Over a period of less than fifty years, which in cultural terms is insignificant, they were to cause a radical alteration in the domestic environment.

9

A strapwork cartouche, copy of a plate engraved by Frans Huys after Hans Vredeman de Vries, (1527-c.1604) The flat bands of scrolled plaster strapwork which Rosso Fiorentino developed for the surrounds of his pictures at Fontainebleau were to become, in Northern hands, the almost metallic scrollwork shown above. De Vries has been held to have set the style of decorative strapwork for the Protestant world. (Victoria and Albert Museum).

Strapwork used as a frieze in oak panelling. This particular style, using opposite pairs of "S" scrolls, with acanthus-like foliage used to fill up the rest of the rectangles, was in vogue for perhaps forty years. It appears on fixed woodwork in churches and houses, and on many items of oak furniture.

Although carving on late Elizabethen furniture could be profuse, and thus confusing, there are probably only half a dozen basic motifs which appeared in one form or another on table boxes and other furniture. Before considering these individually however, it is worthwhile to briefly examine the manner in which they made their way to England.

In the early 16th century, some of the great artists of the Italian renaissance such as Raphael and Rosso Fiorentino studied and imitated the "grotesque" style of decoration, so named from the grottos below ground in which they had been found. Raphael in particular used the grotesque so successfully that he developed it into a complete style in its own right. Paintings incorporating the "new" images such as those of Raphael which decorate the Vatican loggia became a benchmark for study by scholars and architects for many succeeding generations; indeed they appear to have been more studied than the originals.

Through the medium of engraving Raphael's interpretation of the grotesque soon became common throughout the whole of Europe, and although the pictorial aspect of the grotesque need not concern us, the bandwork used to enclose and strengthen the elements of the designs, as an aid to composition, was to develop into "strapwork" which is one of the commonest motifs to appear on table boxes. It is perhaps the only design to have been radically changed during its trip through Italy. Rosso Fiorentino, who was more prepared than other artists to distort classical emblems, came to adopt the most intricate strapwork patterns to surround his paintings, adapting one of the most ancient of motifs, the scroll or volute into a new type of ornament which eventually would appear in profusion on buildings and furniture throughout northern Europe.

Italian influence upon Renaissance ornament in England was profound, beginning perhaps with the design of the Henry V11 monument in Westminster Abbey by Pietro Torrigiano in 1519. In the following reigns, other Italian artists such as John of Padua and Toto dell' Nunziata, and architects such as Sebastiano Serlio brought their influence to bear, mostly working under the patronage of the Court.

Architectural tracts, such as Serlio's "Sette Libri d'Architectura", published in 1556, circulated widely and made a decisive contribution to the spread of the "new" styles of building and decoration, not only in England, but in France, Germany, and the Netherlands. The first English translation of "Architectura" did not appear until 1611, and this in itself was a translation from the Dutch, but the Italian original, in six books, had been circulating in England for many years previously.

This flow of ideas and images from south to north was sometimes direct in that Italian artists and artisans actually worked in England, but there were also indirect influences at work, drawing their inspiration from the same source, but via other northern European countries.

During the sixteenth century Protestant refugees from France and the Netherlands settled here in great numbers; many of these were skilled artisans already familiar with the decorative designs of the French and Flemish Renaissance. Hans Vredeman de Vries, who died in 1604, had worked in Germany, Antwerp and Amsterdam and had produced pattern books which contained most of the popular decorative motifs used in England towards the end of the sixteenth century. His "Differents Pourtraicts de la Menuiserie" was one of the earliest pattern books illustrating furniture published in northern Europe and was widely copied. His earlier work had been greatly influenced by Serlio, but his son Paul was publishing furniture designs as late as 1630 which also contained patterns for strapwork.

The late 16th/early 17th century strapwork may be held to have been a mongrel designed by the Italians but there is another element in its development which is of importance.

Most strapwork on furniture is based on the "S" curve, which has the most ancient origin. The "S" scroll based on plant stems can be found in one form or another in most of the ancient countries surrounding the Mediterranean, and it was to arrive in Venice in about 1530 at more or less the same time as Rosso Fiorentino was developing his own interpretation of scrollwork for architectural embellishment. One of his assistants who also worked with him on the decoration of the Palace of Fontainbleau was Francesco Pelligrini, who wrote one of the first treatises on Moorish designs. Many of these were based on the most intricate interlacing of scrolled plant-stem motifs, and although there is no direct evidence that Rosso was influenced by the introduction of a "natural" element into the formal strapwork bands, it cannot be coincidental that eventually most strapwork incorporated foliar elements in the designs. The ends of the "S" curves on the box which heads this chapter have grown leaves at the ends of the scrolls, and this is typical.

The original Graeco-Roman forms of the plant-stem followed the same "S" configuration, but were further ornamented by the use of elements of the acanthus leaf, and here again this motif would form a design that would appear again and again on buildings and furniture at the end of the sixteenth century, lasting until well into the seventeenth. Due to the possibility of repeating patterns based on the "S" curve it was extensively used as decoration in one form or another on desks and table boxes. It was equally well suited for friezes on larger items of furniture such as beds and court cupboards, and for the rails of tables.

This enthusiastic use of a completely exotic plant form is typical of the Elizabethans who seem to have had an unquenchable thirst for the bizarre. The plant is indigenous to the Mediterranean and at the time very few northern Europeans can have actually seen it; its use is paralleled by the carving of chrysanthemums later in the seventeenth century. Cultivated varieties of this flower did not reach Europe until 1789, and the carvers can only have copied the motif from imported far eastern pottery or lacquer work.

Paired "S" scrolls on the front panel of a table box of the early 17th century. In this configuration they have formed a strapwork pattern. Elongated, they would form the same design as the "S" curves on the box illustrated at the beginning of this chapter. Note the punched decoration within the bands; this was a quick and easy method of embellishment which appears on many desks and boxes during the early part of the century. These scrolls also demonstrate the effect which the carver could achieve with few tools. The body of each scroll was cut with a "V" tool, and the other elements in the design were made with only one gouge of perhaps 3/4" diameter. I suspect that a template was used to outline the scrolls.

In ancient Greece, the "S" scroll was also used on columns of the Ionic order; this "volute" form was eventually to find extensive use on furniture, not only as decoration on panels but also on chairs as arm rests, and on tables as feet. For these latter purposes, its use extends to the present day. As the shape of the volute could be extended at will, it also found favour as a decoration for 17th century panelled rooms.

I believe that one reason for the popularity and longevity of the "S" scroll is that it is easy to make, given that even the most primitive human hand could pivot at the wrist. The shape also seems to fill an oblong in a most satisfactory way; it is also a "natural" shape in that it is found in many plant forms.

"S" scrolls, whether used in their plain form, in combination as strapwork, or as part of a design incorporating other motifs, are probably the most used of all the motifs from the ancient world on furniture of the late 16th and early 17th centuries.

Box no. 2.   Late 16th/early 17th century.

The bold scrolling on this box is unmistakably Greek in origin; on this specimen the side panels are similarly carved. Beneath the lock-plate is a foliar band which separates the two panels bearing volutes; these are mirror-images of each other. The foliar band could also be a Renaissance derivative, perhaps an attempt at the ancient "stiff-leaf" motif.

To the original owner of the box the four roses could have had many meanings. Carvings of the wild English dog rose appear on many articles of furniture and are perhaps the first indication of the Anglicisation of Renaissance designs. The most obvious interpretation is that their presence is political as well as decorative, representing either or both of the roses of the Houses of Lancaster and York.

14

17th century frieze above panelling from Upton Court, Herefordshire.

The rose, however, has sometimes been associated with secrecy; perhaps for this reason many Confessionals were decorated with them in the late 16th century. On the other hand, in Christian iconography the rose has always been a symbol of purity and innocence; the Virgin Mary is usually depicted with them.

In Greek mythology, the rose was associated with Venus.

Today, symbols do not have the great importance to us that they had to the people of the sixteenth and seventeenth centuries; indeed it is difficult to think of any twentieth century domestic symbol of consequence. Politically symbols such as the Swastika have affected the lives of millions of people, mostly adversely, but in decorative terms nothing in our present culture holds any significance to the vast majority of people.

Most of our present-day symbols identify possessions or products; "M.G." identifies a car, "BP" a petrol, "Daz" a washing powder. By contrast symbols in the sixteenth century defined ideas.

The decoration on this very early box must have been a daily reminder to the original owner of something, but from our far-distant viewpoint we can only guess as to what that might have been. If I am right in assuming that many of these table boxes were made to commemorate marriages (and many bear two sets of initials carved with the original design), the presence of the roses has an obvious meaning, but they could equally signify the long period of peace in England following the Wars of the Roses, and the resulting prosperity which had enabled the original owner to buy it, and indeed to have a need for it.

For an early box it is quite large, at 26" long, 21" wide, and 10" deep. For this depth of box to become common we must wait until the end of the next century.

That it is a "country" piece is fairly obvious, particularly if the standard of carving is compared with the paired scrolls on the previous page. The roses on the front panel are deeply undercut beneath, which gives them a sculptural quality; this they share in common with the next box illustrated.

Box no. 3   Late 16th/early 17th century.

The plant-stem decoration on this box has an "Arabesque" source; although it is still a Renaissance motif it also relies on the "S" curve as a basic feature.

Initially, we probably need look no further than Venice to discover the route by which it came into Europe, simple plant-stem designs formed the basis for the most intricate patterns which appeared mainly on brasswork imported into the city, or produced by Moorish craftsmen who had settled there in the sixteenth century. The original source was probably Damascus.

As a single motif it was very soon recognised as being ideal for needlework, and pattern books for embroidery such as N. Zoppino's "Esemplario di Lavori" began circulating as early as 1529.

In England, floral decoration as an embroidery motif flourished during the reign of Elizabeth, and towards the end of the century was not only practiced by the ladies of the Court but had spread into the leisured merchant and yeoman classes; embroidery was to remain popular as a pastime well into the 17th century.

The woodcut drawings in Gerard's "Herbal" which was published in 1597 were used as patterns for embroidery; Richard Shoreleyker's "The Schole House of the Needle" published in 1624 provided patterns for flowers such as the pansy and the wild rose.

This appearance of what is probably a needlework design on the front of a box is interesting in that it may give us a clue to one of their uses. Embroidery materials were expensive, especially at a time when silver-gilt and silver threads were used; these would surely need to have been kept secure. This possible link with embroidery would also confirm that many of these boxes were for the use of the women of the household.

The carving of the roses on the front of this box is very similar in technique to those on the previous box; they are deeply undercut. However there is a difference in the design of the flowers which may have been intentional; perhaps again the to symbolise the roses of York and Lancaster, or perhaps to illustrate the roses of martyrdom or piety, which traditionally were either red or white and would be carved differently.

16

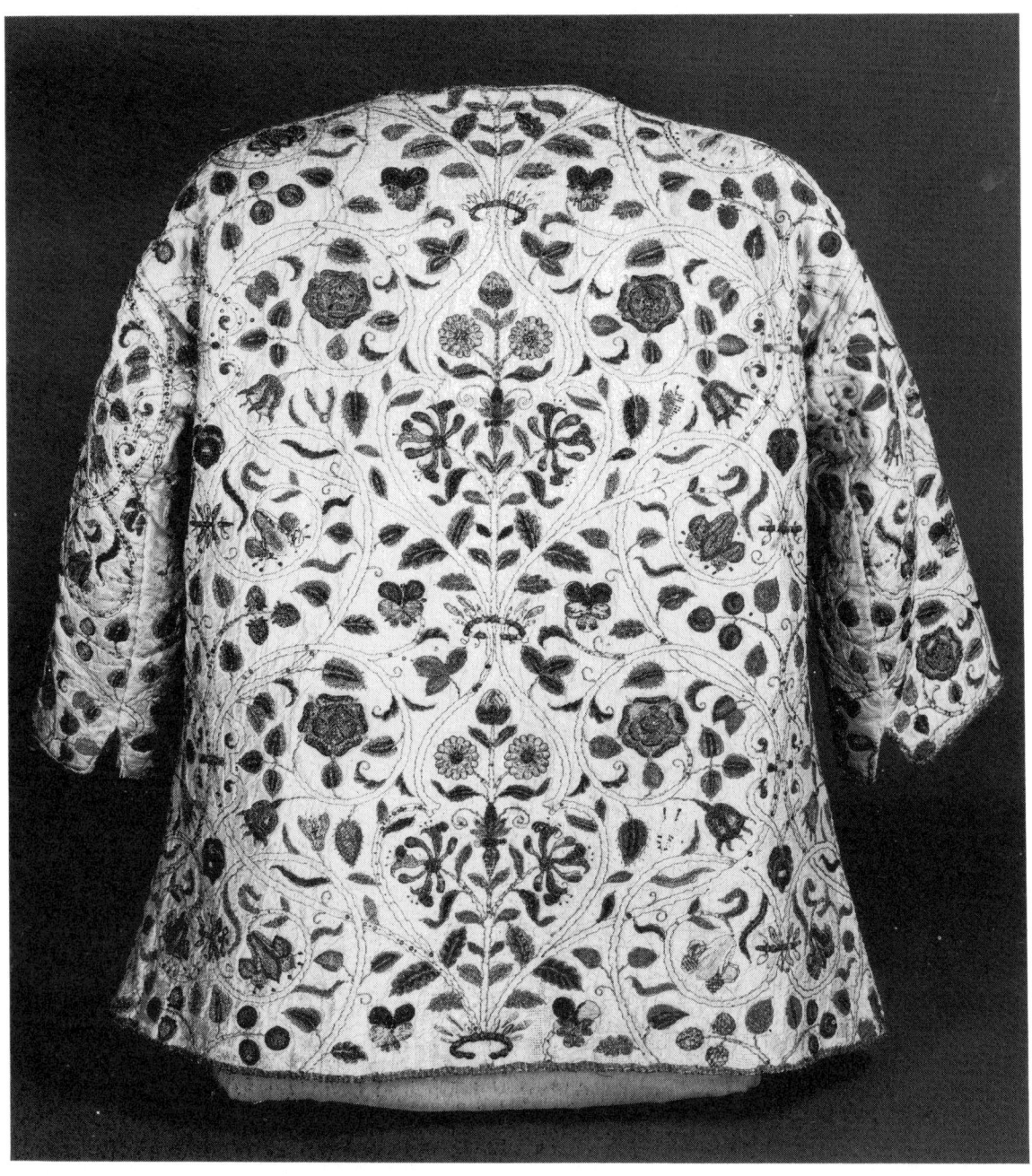

Elizabethen woman's jacket, c. 1600. (Victoria and Albert Museum Collection). Silk, embroidered with coloured silks, with silver-gilt and silver threads. The overall design includes plant stems in a pattern not unlike strapwork; the individual roses spring from the inner curves of these in a similar manner to the roses on the box.

Box no. 4.
Late 16th/early 17th century.

18

This handsome desk carries one of the most ancient plant-stem motifs, and the carver again utilised the "S" curve to fill the oblong spaces. The honeysuckle featured here is not our indigenous variety but the Greek anthemion which was one of the most popular motifs of classical ornament. It appeared in classical architecture as a feature on Ionic capitols or as a "filler" for friezes and architraves and was also enthusiastically employed during the Renaissance. It is an obvious choice for "running" patterns along borders; here it admirably fills the somewhat narrow front panel of the desk. The "S" curve allows it to appear in the fashion favoured by the Renaissance, with one bloom facing upwards and the other down.

 The desk has a further Renaissance feature; each side panel carries a large carved rose, with fleur-de-lis spandrels. Another early feature of the desk is the holly and bog-oak chevron strip which runs around the honeysuckle panels; this also is used on the side. This form of inlay is not often found on the desks and table boxes of the period, although it was extensively used on larger pieces of furniture.

Box no. 5   Late 16th/early 17th century.

This little box carries a very similar honeysuckle motif, but in addition has been profusely embellished by punch-decoration. The margins around the front panel carry alternating chevrons achieved by simply driving a chisel-point directly into the surface of the oak; the design is further enlivened by the use of a small oblong punch in a fairly haphazard manner.

The sides are uncarved, but the lid carries the initials "A.E." in Lombardic lettering. The top also has the margins decorated by punch-marks, in addition, it has scratch-block moulding along all four edges in the very early shallow half-round section, which seems to deepen the farther one gets into the 17th century.

"Scratch-block" moulding appears on very many early boxes. It was achieved by filing the required profile onto the edge of a piece of metal and simply passing this to and fro along the surface to be decorated, gradually deepening the cut. On early boxes the resultant flattened half-round section was applied on the top surface of the board as above, but it was also applied using the scratch-block at a slight angle. It is also present as a groove of the same profile. (Please refer to page 136 for illustrations of moulding sections).

The little linenfold box illustrated on page 8 has two vertical scratch-block mouldings of the convex profile on each side of the main panel.

This moulding is worth looking out for as it can indicate an early date, particularly if in conjunction with classical motifs.

This brings us to the difficult problem of assigning dates of manufacture to these early boxes. Strapwork patterns were used over a very long period on an enormous variety of plain surfaces not only on furniture but on and in buildings. Very few early boxes were dated.

Even when a firm date can be put to a piece of carving this is no guarantee that this was the first date when it was employed: all that we can say is that at that date the design was used. Thus although the strapwork motif on the front panel of box number one (page 9) is virtually identical with that on a pulpit at Wickington church in Norfolk, which probably dates from the 1630's, it also occurs on fixed woodwork elsewhere at an earlier date.

It also does not seem possible to detect a developing style either in the designs or the carving until much later in the 17th century.

For me, part of the fascination of these early boxes lies in this dilemma, and I wonder sometimes whether we are far too cautious in attributing possible dates of construction to them. Strapwork patterns were perhaps circulating in England at least from the 1570's and given the prosperity of even the "middle" classes at that time there must have been a good market for the boxes. Given also the thirst for the "new" there seems no reason why many of the boxes which we cautiously assign to the years of James Ist should not have been made in the last twenty or thirty years of the previous reign.

Box no. 6.   Early 17th century.

The design of the front panel of this box, with its bold, confident paired scrolls has such a "High Renaissance" appearance that it is difficult not to give it an early date on grounds of style alone. A safe attribution however cannot rely entirely on style , particularly as there appeared to be a second burst of flamboyance in design in the 1630's.

A possible clue as to an earlier date lies in the lockplate; this is almost certainly the second lock fitted to the box. Many boxes are found with evidence of replaced locks; on this box this lies in a darker patch on the oak just to the left of the existing plate, hinting that an earlier, slightly wider lock, was originally fitted.

21

Most of these early locks were quite simple mechanisms and probably gave good service for many years, although in the early years they must have been locked and unlocked continuously. Most of the locks which we encounter on desks and boxes today still work, if one can find a key. Locks must have been replaced either when they simply wore out, or the owner lost the key. Just how long a lock would last is a matter of conjecture, but one can find many boxes which are into their third. An average life of forty to fifty years does not sound unreasonable.

The lock on box number six could therefore have replaced the original in 1650, when this style of lockplate seems to have been available, putting a "perhaps" date to the box somewhere around the first ten years of the 17th century.

Two other factors seem to indicate an early date. The lid of the box carries along the front edge the shallow half-round moulding already mentioned; in addition it is very robust. The oak from which it was made averages 3/4" in thickness.

The next three boxes are all "dated", in that at some time, for some reason, someone has added a date.

Box no. 7   Early 17th century

When I first saw this desk I wondered whether it represented a classical revival towards the end of the century; it was firmly dated 1681 on the lid. On closer examination however several things became obvious, the most important being that I could not match the gouge-diameters from the carving on the front panel with those on the lid. Secondly, beneath the "M.W." initials on the lid were traces of earlier lettering of the initials "A.P". These had originally stood proud of the background; they were removed and the newer initials carved into the ground. Thirdly, it became obvious that the borders around the new initials had at some time been painted black; the whole thing was a funereal device commemorating a death.

22

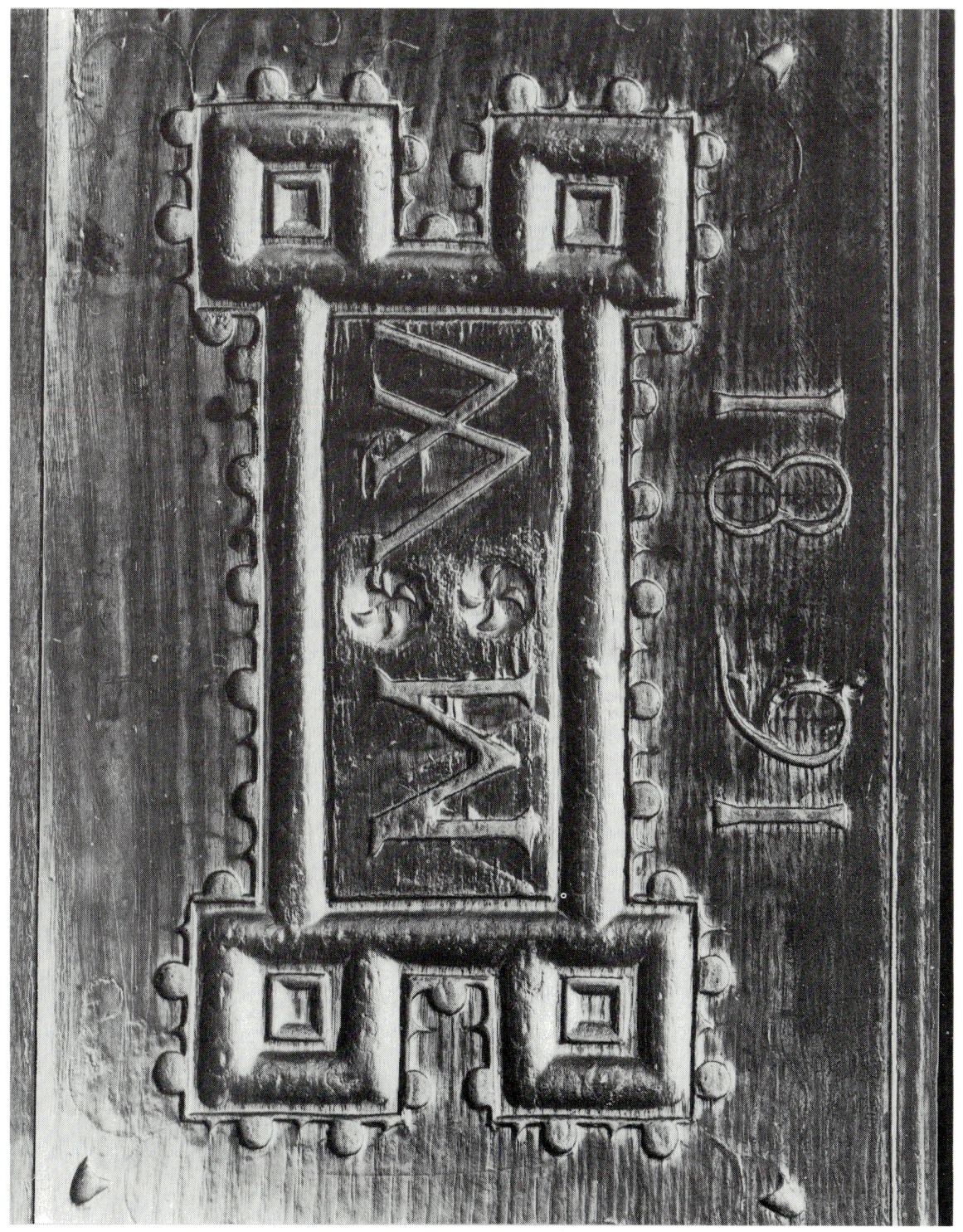

Detail, lid of box no. 7. The moulding on the right of the photograph is a shallow "ogee" in section which might indicate a later date for the lid itself. The two "footballs" in the middle of the initials are from the original carving and are a motif from earlier in the century.

23

Detail, front panel of box no.7. More "punchwork" decoration, made with the tip of a small gouge driven vertically into the bands of the strapwork. Turn a pair of these "S" curves sideways and elongate them and you again have a similar design to that on box no.1 (Page 9).

Once all these factors had fallen into place it was obvious that one had to rely on the classical motifs to date the desk, at perhaps seventy years earlier. The atmosphere of the design is very "arabesque" and it can be no coincidence that it matches almost precisely the detail of these mosaic tiles from the Italian Cathedral at Monreale.

The lock-plate on this desk is the original and is ornamented, with each side crudely copying the shape of late 16th century "cock's head" hinges. Inside it has three small drawers tucked up beneath the "roof", each with an original yew wood knob. (It is rare to find a desk with these still all present.) These might more properly be described as "tills" as they are too small to contain anything other than cash or jewelry. Larger items such as letters and invoices would have been kept in the desk itself.

24

Box no. 8..   Early 17th century.

This little desk has the date "1668" inscribed on the front panel just above the lock, to the right of the hasp. (Unfortunately in the photograph it is hidden in the shadow of the overhanging lid.) The lock is I believe the third, and with a brass cover to the keyhole is probably Victorian at the latest, proof I suppose that it was still in use during the nineteenth century and had to be locked.

The strapwork however is firmly seventeenth century and, I believe, early. Running along the whole length of the flat board on top of the desk from which the lid hinges, is a double scratch-block moulding; one channel is convex, and the other concave in section.

The initials "I.C." beneath the lockplate are original The date is merely stamped into the surface of the wood, using a straight chisel point for the "1" and semi-circular marks from the tip of a gouge to form the curves of the other figures.

A positive clue to dating lies in the decoration of the side panels, which include a very Renaissance "guilloche" motif, encircling a rose. This is almost identical to a similar device which decorates a pulpit-front in the Victoria and Albert Museum, which carries a genuine date of 1628.

Within the guilloche bands, the desk carries punch decoration which is almost identical with that within the "S"- curve bands on the previous box.

25

Box no.9.   Early 17th century.

The strapwork design on this desk is virtually the same as that on the previous example, except that the paired "S" scrolls are doubled. The most interesting feature is the date which someone at some time has taken considerable trouble to place on the front panel. It occupies the space left when the original lock, for some reason, was removed.

When trying to determine the history of a desk such as this, in the absence of hard evidence one usually has to make assumptions. If we assume that the date of 1662 does represent the date when the alteration was made then the strapwork design was probably carved sometime in the 1620's, and this seems reasonable.

Originally, the initials "O.G." were part of the earlier carving; the inserted piece of oak just touches the top of the "G". The carver of "1662" went so far as to imitate the style of the lettering, running a "V" tool along the numbers to match the groove cut in the letters. He also fashioned the two semi-circles under the top margin to match others elsewhere in the design of the front panel. He went further; the piece of oak which he let into the lock space has a fine grain which almost matches that of the rest of the panel, and finally he chose a pointed matting punch of virtually the same size as that used to matt the ground behind the main features of the original design. So, was all this done in 1662? We have no way of knowing; the date may have been demanded by an owner for some other reason, perhaps to celebrate a birth.

One further mystery surrounds the desk; the lid is not the original. It has a chip-carved margin of an indeterminate design, and the hinges are of brass. There is evidence on the rim that the hasp belonging to this lid would have fallen to cover the left hand "6" in the date, but on the top of the front panel is a slot cut to accommodate the original, the hasp of which would have come somewhere between the second "6" and the "2", i.e. the keyhole would have been to the left on the original lock. All the lockplates that I have ever seen on boxes and desks have the keyhole on the right. Another assumption: the original was fitted upside down.

26

There is a further clue to the age of the desk. The replaced lid hinges from the old top board. This is of the fairly standard width of 6" (which I believe was to accommodate a candle-holder) and bears the now familiar flattened half-round scratch-block moulding, even on the sides. This does not match the lid.

Unlike box no. 7, this one has lost its drawers, although the shelf which carried them is still in place.

Originally this desk was integral with a stand; the tops of the legs of this still exist inside the desk at the corners. These are of beech, and one is tempted to think that this is why the stand was amputated; beech is very attractive to woodworm.

Box no. 10.   2nd quarter 17th century.

This little box is another showing a combination of a classical motif with scratch-block moulding, plus a replaced lock. The foliar tips of the "S" scrolls have become acanthus leaves, but there is an absence of gouge work in the carving. The execution of the design begins to approach that of later in the century, where motifs were simply outlined with a "V" tool and the ground was seldom excavated to a depth of more than 1/8". We have at last arrived at a "country" piece, with all the hazards that that represents in dating.

We know that away from the bigger centres of population the box makers (who were very often the village carpenter) carried on reproducing designs long after they had been superseded in the towns. This conservatism perhaps lay not in the makers but in their customers; there was no point in producing something which did not sell. If acanthus leaves had always been the flavour of the month, there was little point in trying to sell a carved tulip to a farmer, who had never seen one anyway. Also, in dealing with country workshops we have to be careful not to attach too much importance to the scratch-block moulding as an early indicator of date; no doubt they all had them. They were simple to make on the premises, easy to use, and quick. (An educational project for a wet Sunday afternoon is to file a profile into an old steel cabinet scraper and try it out on a piece of seasoned oak. An amateur can decorate the lid of a box in ten minutes).

Good tools in the 17th century were expensive and very often were passed on from father to son; witness their frequent mention in probate inventories. Moulding planes were relatively sophisticated items and there must have been many country workshops which did without them, relying on the tried and tested scratch-block instead.

Apart from the matting of the ground, there is very little punchwork decoration on the front panel of this box; all that we have is a series of small holes put in with a round, possibly pointed tool on the raised surfaces of the scrolls.

As to dating the box, one can only guess, but I would put it somewhere in the later years of the reign of Charles Ist. If one accepts a country provenance, then it could be held to be in the style of that period.

28

Box no. 11.   2nd quarter 17th century.

Stylistically this little desk has much in common with box no. 10 on the preceding page. The acanthus terminals to the scrolls have changed into vine leaves, leaving me to interpret the devices on each side of the lockplate as bunches of grapes. To accurately depict grapes in wood-carving takes some skill, and implies the possession of the correct tools; I suggest that the carver of this front panel had neither. Indeed, the inside surface of the panel bears a sort of "trial run" of the motif (which has proved impossible to photograph successfully). Had the carved grapes become convex instead of concave the effect would have been more realistic, but what the carver managed to achieve with a single gouge has a certain charm.

Grapes and grape vines have always been a popular decorative motif, and have been used by many cultures, ancient and modern. They were a favourite component of Renaissance decoration, possibly for religious reasons. In Christian iconography grapes represented the Eucharistic wine. On the other hand, in classical mythology they were symbolic of Bacchus the Roman god of wine and fertility. The original owner of this desk, if educated, would have been well aware of both.

Again, this is a country piece, typical of its period, which I suggest is about the same as the last box.

One feature which it shares with all the preceding examples is the size of the nails which join the various boards from which it is constructed. The heads of most of these average 1/4" diameter, unlike most of those used in boxes of the late 17th century, which are far smaller. The nails for these country boxes would have been made locally and the local blacksmiths were probably unable, even had they wished, to make nails of smaller dimensions. And why should they? Finer nails were not needed until more sophisticated furniture was made at the end of the century.

Box no. 12.   2nd quarter 17th century.

The carving on the front panel of this desk carries a much more elaborate pair of "S" scrolls than we have hitherto seen; the forms appear to be a hybrid between the volute and the grotesque. On each side of the lockplate they terminate in a most ugly pair of heads, which point upwards. In addition, as if these were not enough to frighten the children (perhaps their purpose) on each of the side panels the carver has placed a most ferocious dragon.

Dragons were a favourite motif in woodcarving and crop up on all sorts of pieces of furniture, and although they are the national emblem of Wales their employment was widespread elsewhere, particularly in the Westcountry.

Dragons have a most ancient pedigree; the name is derived from the Greek word for "snake". They have always, in Western mythology, been deemed evil, but at the same time they were thought to be the guardians of treasure. Could this be the reason for putting them on this desk?

In Chinese mythology, they are particularly ancient, appearing on bronzes from the end of the 2nd millenium B.C. By medieval times they had swarmed into England, appearing in illuminated ecclesiastical manuscripts, and on church woodwork. Their route into Europe is thought to be from designs imported on silk from the east. In the course of their travels they changed their character; their Chinese ancestors were benign creatures who could bring rain when needed. They were also protectors of the innocent, and, again, guardians of treasure. In China the dragon was the emblem of the Emperor; in England it obtained Royal significance from Anglo-Saxon times. Richard 1st took it as a battle standard on the First Crusade.

A further element in dragon mythology is the Chinese connection between dragons and pearls. The legendary pearl was thought to represent the sun, which the dragon chased across the sky but never caught. Within the bands of the volute form on the front of

Pair of carved dragons on Gloucestershire chair.

Side panel, box no. 12.

Dragon from "Serpent and Dragon History". Book 11, 1540.
Copied by Edward Topsell in 17th century natural history books.

31

the desk are four spherical dots, with a fifth below the jaw. Could these represent pearls? If so they could represent the achievement of the impossible. This mysterious dot appears on other carved dragons; I found it on the cresting rail of the Gloucestershire chair of about 1630 illustrated on the previous page.

Box no. 13.   2nd quarter 17th century.

There are also pellets on the flanks of these dragons, on a somewhat smaller and less intimidating box. Whereas the dragon on the previous example has a forked tongue, these are breathing fire very enthusiastically and appear to be smiling as they do so. Dragons in this form appear all over the place in 17th century woodwork, particularly in churches.

At some time the lock of this box was removed and the present shield inserted most carefully. There is a small mystery here, because the very bottom of the present shield formed part of the original carving; the inserted piece can be clearly seen in the photograph. Therefore the original lock must have covered the original shield. The initial presence of a lock is confirmed by nail holes beneath the lid which once carried the shaft of the hasp. Perhaps a previous owner decided to improve the appearance of the front panel by replacing the shield. But if he had gone to this trouble, why did he not arrange to have initials carved on it?

As to the date of manufacture, the box may be earlier than the preceding desk. The front edge of the lid carries the early scratch–block moulding, whereas the only moulding on the other dragon box is that at the bottom, which is an eighteenth century replacement.

If it were not for the slightly "country" look to this box I would be tempted to allot it an earlier date, it could well be from the 1630's. There is an element in the construction which makes me think otherwise; the back and two side panels are of elm. If the maker had simply just run out of timber, this would have been readily to hand in a country workshop; alternatively of course it was probably cheaper. I have seen this mix of oak and elm in many chests and boxes, with the elm always in the less visible places.

Box no. 14.   3rd quarter 17th century

At first sight it appears difficult to put a date to the strapwork on this box; in general form it it similar to paired "S" scroll patterns already illustrated. However, if one compares it with that on box no. 1 (page 9) differences become immediately apparent. The carving on the earlier box is elaborate and fussy, on this one it is plain and simple. The terminal foliage on the scrolls of the early box is acanthus–like; on the later it has more in common with that found in association with the stylised tulip heads found on many late-century boxes.

The bands of the scrolls are undecorated and there is a pleasing regularity about the whole design.

The strongest clues to date lie in the construction of the box; the front panel is not nailed onto the side panels but sits between them. Instead of nail heads being visible on the front of the box, these have been driven into the front from the side. This leaves the joint between the panels visible from the front; in slightly later boxes this was to be covered by a strip of half-round moulding. (See the dated box no. 47, page 114).

Another late feature lies in the moulding of the lip of the lid and that along the bottom of the box; these were both made with a plane.

It is perhaps a tribute to strapwork as a design that it lingered on so long, particularly in the northern counties, where it can be found on court cupboards of the 1680's. It was probably, from the maker's point of view, always a "safe" design to employ. It made no political statement and presented an image of security and respectability. It was above all a design borrowed from architecture and during the seventeenth century could be found on buildings throughout the country. It was a particular favourite of the ecclesiastical carvers and could be almost universally found on pulpits and other church furniture.

Given the conservatism of the times, it is not surprising to encounter it so often on desks and table boxes. The box makers were not only craftsmen; they had to sell their wares, and even this late in the century, strapwork on a box would have found a market.

Box no. 15. Late 17th/early 18th century.

This treatment of the ancient "S" scroll looks incredibly modern; Salvador Dali would have found it familiar. The execution of the work is typical of the period, however, in that only two tools were used, a "V" tool to draw the outlines of the main elements and an angled gouge to remove the excavated part of the design. This appears to be standard practice towards the end of the seventeenth century, continuing into the eighteenth, and I find it difficult to understand. Perhaps one should try and consider it in relation to what had happened to other furniture which was "modern" at the time.

Carving into the surface of wood to produce a design became gradually superseded by applying patterns to the surface in the form of veneers or marquetry; this latter coming into the country in the 1670's. Judging by the amount of furniture so decorated which still remains with us today, it must have become immediately fashionable. The unfortunate country craftsmen who lacked the skill to make it could only attempt to adapt their existing techniques to present a greater area of plain surface on their work, simply drawing their motifs with the one tool in their repertoire which could do this.

This oddly decorated box is from Yorkshire, which at the time was about as "country" as you could get. It has some interesting elements in the construction; the front panel extends beyond the side panels for perhaps half an inch on each side. At the top of the panel on each side a slot has been provided to accept the substantial battens which are beneath the lid. These battens form the hinge system; at the back of each a dowel has been driven into the side panels, providing a fulcrum. This arrangement is very often found on country boxes, no doubt to save the cost of metal hinges, but the dowels tended to wear out and one quite often finds boxes where a metal hinge has been added later.

These overlapping front panels are also found on some Westcountry boxes of the seventeenth century.

Box no. 16.   Dated 1711

Despite the elaborate appearance of the design on the front panel of this box, again, it was accomplished with only two tools.

The "S" scrolls are elongated and formalised, and we are a very long way from the volute forms of ancient Greece. This is "sampler" territory, and the roses, date. and initials could equally well have been accomplished in embroidery.

By this date, chests of drawers have taken over many of the functions of the table boxes: this one, I believe, may have been designed as a workbox in which to keep needlework materials. It is an odd size, compared with the previous boxes, 27" long, 8" deep, and only 12" wide. This narrow width makes me think of knitting needles.

The front panel overlaps the sides again, as in the previous box; possibly this too is a late feature.

When the first box in this strapwork sequence was made, the owner probably firmly believed in the existence of the Unicorn, and the Church was insisting that the earth was the centre of the Universe.

One year after this box was made, Newcomen perfected a steam pump for the drainage of mine-shafts, and the Industrial Age was just around the corner.

Between the two boxes, Gallileo made a telescope and observed the planets of Jupiter, John Napier invented logarithms, Francis Bacon wrote his "Advancement of Learning" and the Royal Society was formed.

A dilution of classicism is apparent in these boxes of the late 17th century; the carving appears to have lost its way in the confusion of new ideas.

The peak years for strapwork were probably those of the Jacobean period, when Inigo Jones brought back from Italy the strictly classical forms of Palladio. The richness of ornament of the early examples and the supreme confidence in their design seems to bear this out.

35

Straight strapwork, from the back panel of a settle of about 1620. At some time, possibly during the
Commonwealth, the panels were covered; hence the lack of wear on the higher surfaces of the carving.

"Fluted" decoration from the frieze of an early 17th century court cupboard.

We come now to the second of our ancient decorative motifs, known as "fluting". This can be found in profusion on furniture of the late 16th and early 17th centuries, and even later on country pieces, and thus it is difficult to use as a dating factor. It tends however to be most common on furniture which one would class in general terms as "Jacobean"; i.e. covering the first twenty years or so of the 17th century. This certainly seems to be so on desks and table boxes, to which it is admirably suited as a "running" decoration.

It is also found with an internal reed occupying the bottom part of the groove, in which case it is known as "stopped" fluting.

As a decorative form it is illustrated by Sebastiano Serlio in his pattern books, and thus has a Renaissance pedigree, but its remote ancestry lies in Graeco-Roman architecture. Vitrivius in the first century B.C. held that it was intended to represent the folds in drapery and certainly when viewed  as a decoration to a column it does give that appearance. On the other hand it also appears as a rim decoration on Corinthian cups of the 5th century B.C. where it looks like a miniature railway viaduct. Earlier, it can be found as a decoration on Assyrian temples.

Box no. 17.   Late 16th/early 17th century.

There is a distinctly architectural flavour to these early fluted boxes, the above desk demonstrates this well. The design is simply that of a series of arches supported by pillars; one can easily imagine the whole thing as a miniature building.

This desk has all the early features already mentioned, a classical motif, profuse punchwork, and the correct moulding which runs all around the lid and even decorates the batten along the lid which stops books from sliding off. In addition it has a "Tudor" lozenge on each side panel, with additional embellishments.

There is an interesting difference between the fluting on the left hand side compared to that on the right. The "pillars" on the left have a vertical channel cut out of them which only extends about half-way up, on the right this goes all the way to the top. Perhaps the carver made a mistake, but I think a more likely explanation is that we are looking at evidence that quite long "runs" of carving were made, and simply cross-cut to the lengths required. The problems of the left and right margins are solved if the carver left enough room for a lock plate between each section of carving, (in this case, seven arches). He could then cross-cut to provide the margins in the lock plate spaces, leaving himself with a small amount of waste wood. On this front panel he was unlucky in that having cut several sections of one style he was left with a panel showing both the old, and a revised style. Rather than waste a complete panel, he used it, and no doubt the first owner didn't notice, or was not concerned.

Eccentricities in design and execution proliferate in early carved work, often lock plates are simply nailed over the decoration in a most untidy manner. This occurs so often that one must assume that it did not matter.

The profuse punchwork on this desk is worthy of close examination; it was almost all carried out with a single punch. (See enlarged detail on page 149).

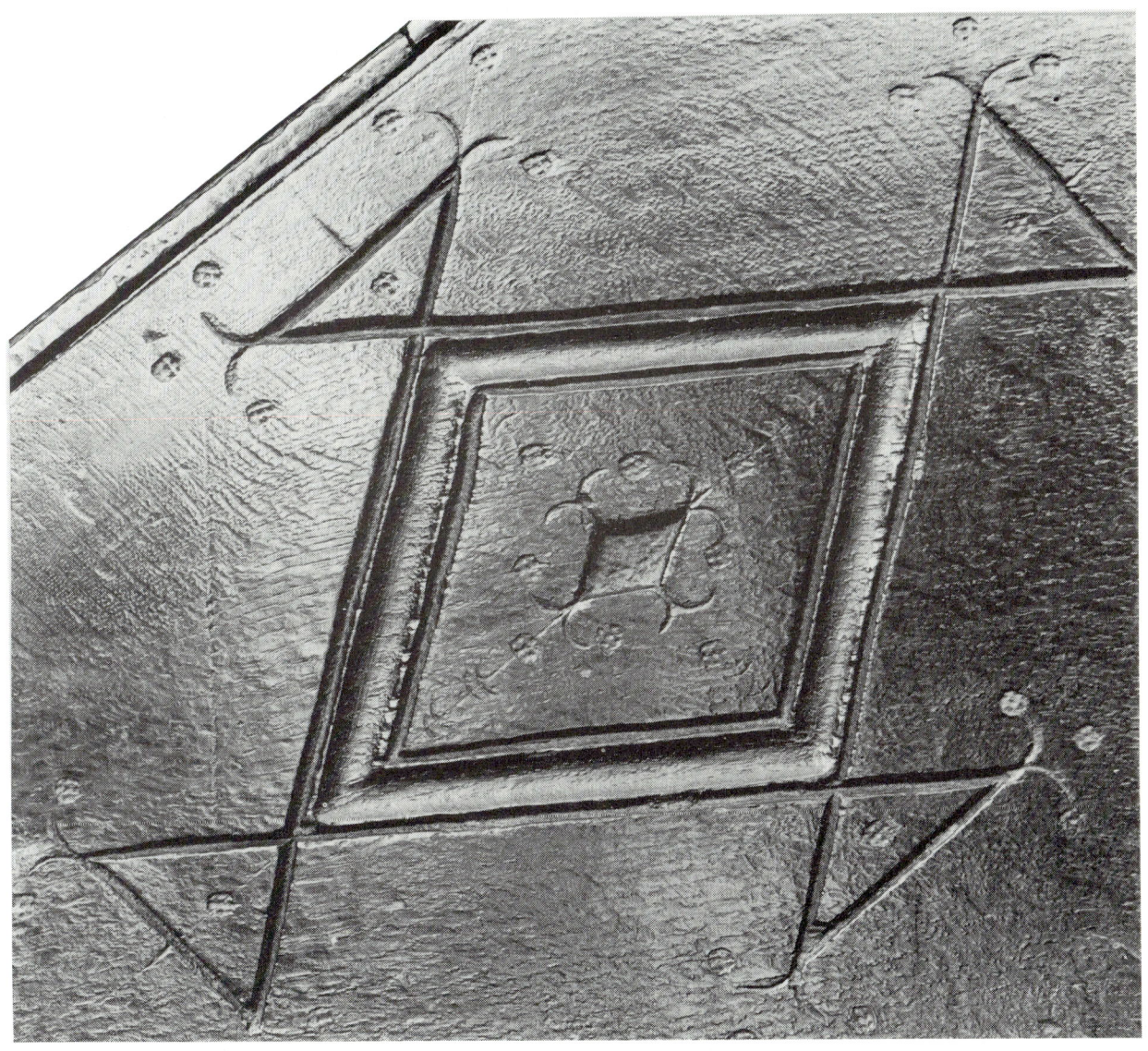

The enlarged detail above clearly shows the shape of the small round punch which made all the "floral" decoration on the desk, which is found within the fluting, along the top of the front panel, on the lid, and the sides. Above the fluting it is combined with the use of the tip of a small gouge which created the effect of leaves on each side of a stem. The original scribe marks which laid out the design are also visible on this panel and have been incorporated into the stems of the flowers.

For the heavier parts of the design, the carver used only three tools, a "V" tool to set it out, and gouges of two sizes to excavate the fluting and the lozenges on the side panels.

The desk illustrates well what could be achieved by a skilled carver using the most basic of tools.

Box no. 18.   Late 16th/early 17th century.

The architectural appearance of this box is enhanced by the precise carving and the deep bottom moulding; it even looks like a temple. This is "stopped fluting," with an internal reed, and it appears extensively on other furniture of the period. It also occurs on church panelling; dated examples exist placing it at the end of the 16th century.

The initials "E.R." cut into the front panel are partially covered by the lock plate, which is an eighteenth century replacement. The initials themselves contain the remains of a black wax which at some time was melted into the grooves for emphasis, this must have been done before the lock was replaced.

This is a large box (31" long, 20" wide, and 9½" deep) and must have been made for a person of some importance bearing in mind the extremely high quality of the carving. The oak from which it was made has a particularly fine grain, and the boards are of substantial thickness.

The bottom moulding similarly is of a high quality, and in section is typical of early work. It is almost identical with that along the bottom of the early strapwork box no. 1 (page 9) These bottom mouldings with their complicated sections can only have been produced by a moulding plane, but the lid moulding on the above box is scratch-block work.

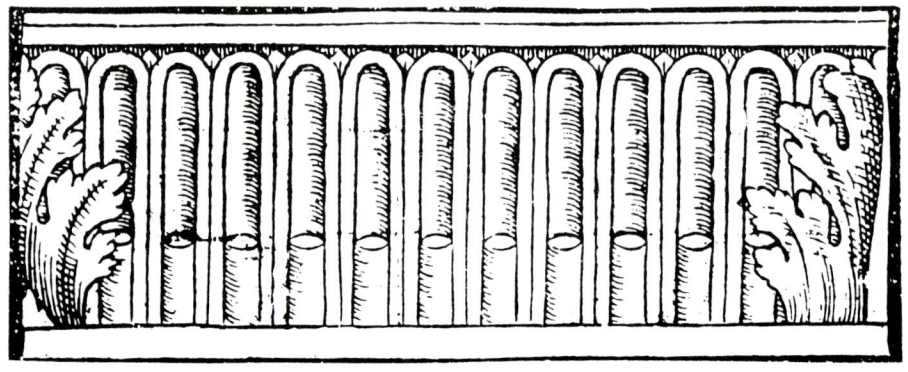

Stopped fluting illustrated by Sebastiano Serlio, in the English translation of "Architettura"

40

Detail of carving on box no. 18, approx. 70% full size. The fine grain of the oak is apparent in this photograph. Many front panels are of oak which has obviously been selected by the carver for its higher quality, and differ from oak found elsewhere in a box.

Box no.19.   Late 16th/early 17th century.

I would have given this little box a later date if it had showed evidence of "country" make but on closer inspection the carving appeared of quite a high quality. At 20" long, 12" wide, and just under 9" deep it obviously contained items of far less value or importance than the box on the previous page,. but it has the same early features in the classical origin of the design, and the moulding along the back and the front of the lid.

Here again the carver chose a fine-grained panel for the carving of the front. There is little punchwork, but the inner surfaces of the fluting were decorated with "V" shaped marks which appear to have been made with the tip of the same gouge which took out the wood from the main body of the design. The small round punch which was also employed has only four segments.

At some time, the lock was removed and the resultant hole plugged with a new piece of oak; this again was chosen to match the existing timber and was carved with a continuation of the original fluting. This was done with equal skill by the second carver.

Before all this happened, the box had had two hasps and a second set of hinges. Whether these antedate the insertion of the wooden plug we do not know, but it is likely bearing in mind the age of the box.

The initials "E.F." are incised into the bottom margin of the panel.

42

Detail; front panel of box no.19.

Fluting on rim of Corinthian Band-cup, 5th century B.C.   (Martin von Wagner Museum, Wurzburg.)

43

Box no. 20.   Late 16th/early 17th century.

This imposing desk is quite large, at 23" long, 20" wide, and 15½" tall: it has two shelves inside which do not appear ever to have had drawers. It exhibits all the early features already illustrated, each side panel bears the same bold fluting as the front.

At one time it had a bottom moulding, which has disappeared. The lid has scratch block moulding around all the sides, this also appears on the batten at the front.

The carver has again used the tip of quite a wide gouge to mark out the little balloon shaped motifs within the fluting; these are repeated on the sides.

The various panels which make up the desk are secured by nails with heads of generous proportions; these can be seen on the left hand side of the enlargement on the facing page.

Detail, front panel of box no. 20.

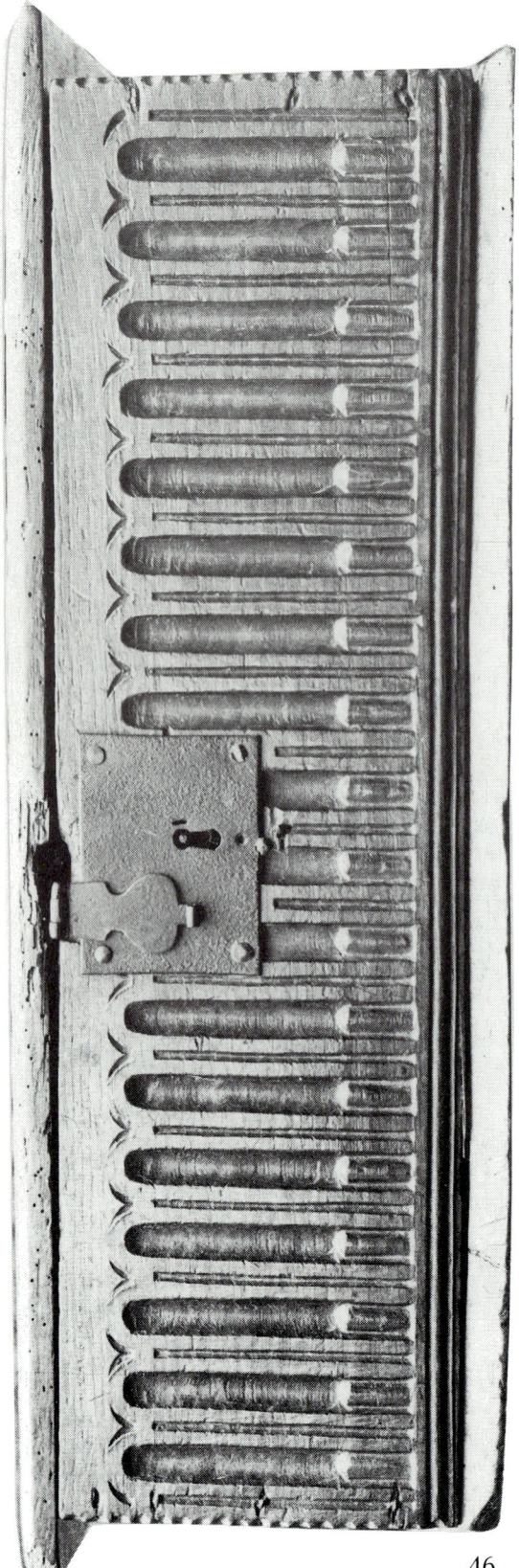

Box no. 21.   Early 17th century

The absence of moulding along the bottom of this box makes me consider that it came from a workshop without a moulding plane. All that "finishes" the bottom of the box is the chamfering of the bottom boards and a strip of half round moulding worked along the bottom of the front panel. Hence the slightly later suggested date.

46

More stopped fluting, this time from the sides of box no. 23, page 52 . The presence of this classical motif in conjunction with guilloche banding on the front panel helps to confirm an early date for the desk. There is a series of punchmarks of a floral shape running along the top of the fluting just below the lid. These are not unlike those on box no. 17 on page 38.

Guilloche banding from the back panel of a settle. (See page 36).

Box no. 22   Late 16th/early 17th century.

One of the simplest and yet most effective of all ornamental patterns is the "guilloche", which on the above box is a repeating series of interlacing bands. In the late 16th and early 17th centuries it was extensively used in wood carving and appears on many pieces of furniture, usually enclosing a rose. It was particularly useful for decorating the panels of table boxes and desks due to its "running" nature. It could also be used to enclose other motifs, as on the settle panel on the facing page; the main features of the design are that the bands intertwine and overlap.

It is a Renaissance motif, and appears in the 16th century pattern books of Sebastiano Serlio, but as a design it has a most ancient origin.

In A.D.79 it was on the walls of Pompeii, to which it had travelled from Ancient Greece.

It was also extensively used in Assyrian decoration.

In common with most other "classical" motifs of course, it was no stranger to England even before the Renaissance, which in many cases simply reintroduced designs which had been brought here by the Romans. Very nice guilloche banding can be found in mosaic form on the floor of room 10 in the Roman Villa at Chedworth in Gloucestershire; this however was lying hidden in the ground until 1864.

As an art form, "circular" guilloche has a long history, but in other shapes it appears to be multicultural. It is found in early Islamic decoration; also in Scandanavian, Celtic, and Anglo-Saxon work, some of the patterns achieved by using it are extremely complicated.

In its simplest form it appears to have been impervious to the passing of time. As late as 1729 it was employed as a wall moulding at Chiswick House in London in exactly the same form as that illustrated by Serlio.

Guilloche decoration on a Rhodian plate, second half of 7th century B.C.   (British Museum).

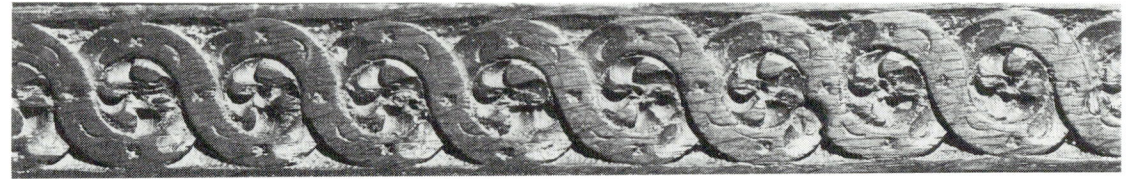

Guilloche carving from the frieze of a 17th century court cupboard.

Guilloche (and fluting) "country-style". From the front panel of a very small coffer made somewhere between 1600 and 1630. Either the compass-work in setting out the design was erratic, or the carver did not have a great deal of control over the gouge. The uneven widths of the fluting might be due to the wear involved in polishing, some of the fluting has a very slight hour-glass shape. Note again the scratch-block moulding along the bottom of the panel. The decorative series of notches on the right-hand edge of the front panel may indicate a Westcountry origin.

Box no. 23.

Late 16th/early 17th century.

Here we have "professional" carving at its best; compare the crisp precision of this guilloche banding with that of the country work on the previous page. There is a pleasing classical elegance in the design of the front panel of this small desk. The sides are decorated by stopped fluting. (See page 47).

Box no. 24.   Late 16th/early 17th century.

In appearance, this is probably the most "Roman" box that I have ever seen. The design belongs to the late Elizabethan period when it was well in favour as an ornament for the friezes of larger pieces of furniture, such as court cupboards. The Victoria and Albert Museum has one which bears this series of alternating squares and circles; this is dated 1610.

The banding is not guilloche; on this box it takes the form of ropework. This can be a Romanesque device; it is found in Norman architecture as a form of moulding. It can also be found occasionally in wood carving associated with Gloucestershire and other western counties. At an earlier date, it was possibly used in imitation of the rope which bound early Christian fonts.

Here, I believe that it may have a nautical significance; the spoked circles which surround the roses give them the appearance of ship's steering wheels, and rope has an obvious connection with the rigging of sailing ships.

The symbols within the squares could well resemble the Roman fasces which were an emblem of authority, particularly of magistrates. One can speculate that perhaps the box was made for the master of a trading vessel, perhaps sailing out of Bristol or some other Westcountry port. As a group, the late Elizabethan merchant seamen had done very well and could easily afford to furnish their houses in style.

It is not impossible that very often these little table boxes must have been carved to order in imitation of carved motifs already present on other pieces of furniture owned by the original buyer. Perhaps this one may have been carved to "go with" a court cupboard or some other larger piece.

Saint Christopher bearing the infant Christ across a river. According to this ancient legend, as a reward for his services, Christ turned his wooden staff into a date-palm. In Christian iconography the palm leaf has always had a deep significance and has been used as a decorative emblem for centuries. (Woodcut, early fifteenth century.)

Box no. 25. Early 17th century.

We come now to a group of boxes where the main decorative feature is the "lunette", an arcade shape where the curves can be struck from a centre by compasses. It is possibly the motif most often encountered on carved oak furniture of the late Tudor/Jacobean period, and was common over perhaps a hundred years of decoration, given that it lingered among country makers long after it had gone out of fashion among the cognoscenti.

This makes the dating of table boxes and desks with lunette panels most difficult, and it is only when other factors such as moulding sections and punchwork are taken into consideration that a box can be termed either early or late. Dated examples exist, but these only confirm that the basic design did not change over a long period.

The spaces within the lunettes carry different motifs, one commonly finds demi-roses with a variety of petals, and leaf forms based on the palm or the acanthus.

In later work a device resembling a whole tree, rising up from wide-spread buttresses can be found. Very often this is in association with an inward curving extension of the lunette bands on each side which calls to mind the shape of the Greek "palmette" which I believe may be the origin of the lunette. In architectural terms, the shape can be found in Norman archways where the curve is similarly struck from one centre point.

In the above box the lunettes have a palmate infilling; the carver has cleverly emphasised the leaf-shape by chamfering each frond on each side. There is a lot of punch decoration within the bands and on the lotus heads which decorate the bottom moulding, mostly carried out with a small "o" shaped punch; in addition, a larger oblong punch was used to matt the deepest parts of the carving within the lunette bands and behind the tips of the leaves between them. The depth and quality of the bottom moulding, and the amount of punch work lead me to give this box an early date. This is reinforced by the fact that the present hinges are the third set to have been fitted to the lid.

55

Box no. 26.   Early 17th century.

The interlacing lunettes on this little box enclose an acanthus–like leaf-form which seems well suited to the lancet shape which the interweaving of the bands has created.

The box has a slightly "country" flavour, but the carving is deep and assured. There is some punch decoration in the deeper parts of the carving behind and above each lunette band, and the whole thing appears to have been executed with only three tools, a "V" gouge, and two curved gouges with a larger and a smaller diameter.

Interlacing lunette bands appear to have a mainly Westcountry provenance, but the device occurs on Norman architecture; it can also occasionally be found on fonts of that period.

The arcading illustrated to the left is from Christ Church, Oxford, which dates from circa 1180 A.D. The pointed arches between the pillars may well have given someone the idea of the shape of windows during the Early English period in Church architecture. On the box, however, with the acanthus association, the whole thing has a very Mediterranean appearance. It must have seemed very exotic to its original owner. For its period, the box is very deep, and this is so unusual that one wonders why. It may well have been made to contain something out of the ordinary.

Detail, front panel of box no. 26.

Box no. 27.   2nd quarter 17th century.

This handsome desk also has interlacing lunette bands, but here the carving is of the highest quality. The infilling of the bands is much more sophisticated and relies on neither the acanthus or palm leaves. There is some foliar carving, but the tips of the leaves are rounded and not notched. The lunette bands themselves are decorated with punchwork in the form of stars and bracket-like symbols.

The lid bears an equally well-carved panel containing a dragon curled up within a pattern of vine-leaves and grapes. He is a European dragon with a serpentine shape and two legs; his tail terminates in one of the vine-leaves. I believe this panel to be a temperance emblem, the dragon is obviously asleep but lurks within the foliage waiting to catch the unwary drinker.

On the front panel the foliar elements within the lunette bands carry the same series of beaded punchwork decoration as those which lie outside the bands on box no. 26.

The carving of the desk appears to have been the work of two craftsmen; there are differences between the matting behind the dragon and that in the lunette panels. Furthermore the style of the carving of the dragon is very like that encountered in ecclesiastic work on misericords, it is certainly of a sufficiently high standard.

The initials on the lid are obviously a later addition, the bar of the "T" cuts into one side of the lozenge. It has been suggested that the second "initial" is a medical symbol, or possibly a representation of the staff of Mercury (but here I would have expected the staff to have had wings).

I think it far more likely that the device is simply a monogram of the letters "E" and "B".

One feature which might make it earlier than I suggest is a scratch-block moulding along the front edge of the lid, but this is the only such moulding on the desk.

58

Detail, lid of box no. 27.

Box no. 28.   2nd quarter 17th century.

The carving on this box is also of a very high quality, again the side panels are decorated with the same lunettes as the front. The lunette bands themselves are flat and carry a simple trailing plant-stem ornament made by driving the tips of two gouges vertically into the wood. The widths of these tools can be clearly seen in the enlargement on the facing page; a wide gouge was used to form the main stem of the ornament, and a smaller one made the side shoots which terminate in a bud. This latter was suggested by the use of a single small round punch.

In the deeper parts of the design between the leaves within the lunette bands the carver has again used a small pointed punch for emphasis; this punchwork is similar to that on box no. 26 (page 57). There is more punch decoration in the shallow grooves cut around the outside of the lunettes; this is similar to that noted in the guilloche bands on the side panels of box no.8 (page 25). This punchwork was particularly easy to make, employing only a round or star shaped punch and a gouge of a small enough diameter to form the "bracket" shapes.

The box is further decorated with a zig-zag strip which runs along the panels beneath the lunettes; this again was a quick design to apply. The margins of the band were laid in with a scribe, and the chevron stripes formed with the tip of a flat chisel driven vertically into the work. This can only have been done before the assembly of the box. This chevron work has a decidedly "heraldic" feel about it, and was quite a popular method of decorating the margins of the lids of boxes and desks, probably because it was quick to make, and effective. One finds it quite often on mid-century pieces, and as a dating factor is well worth looking out for. Occasionally the zig-zag pattern is composed of wavey lines instead of straight ones.

The position of rusted nails along the bottom of the front panel leads me to think that the original moulding was quite deep, and this, together with the amount of punchwork hints at a date nearer to 1630 than 1650.

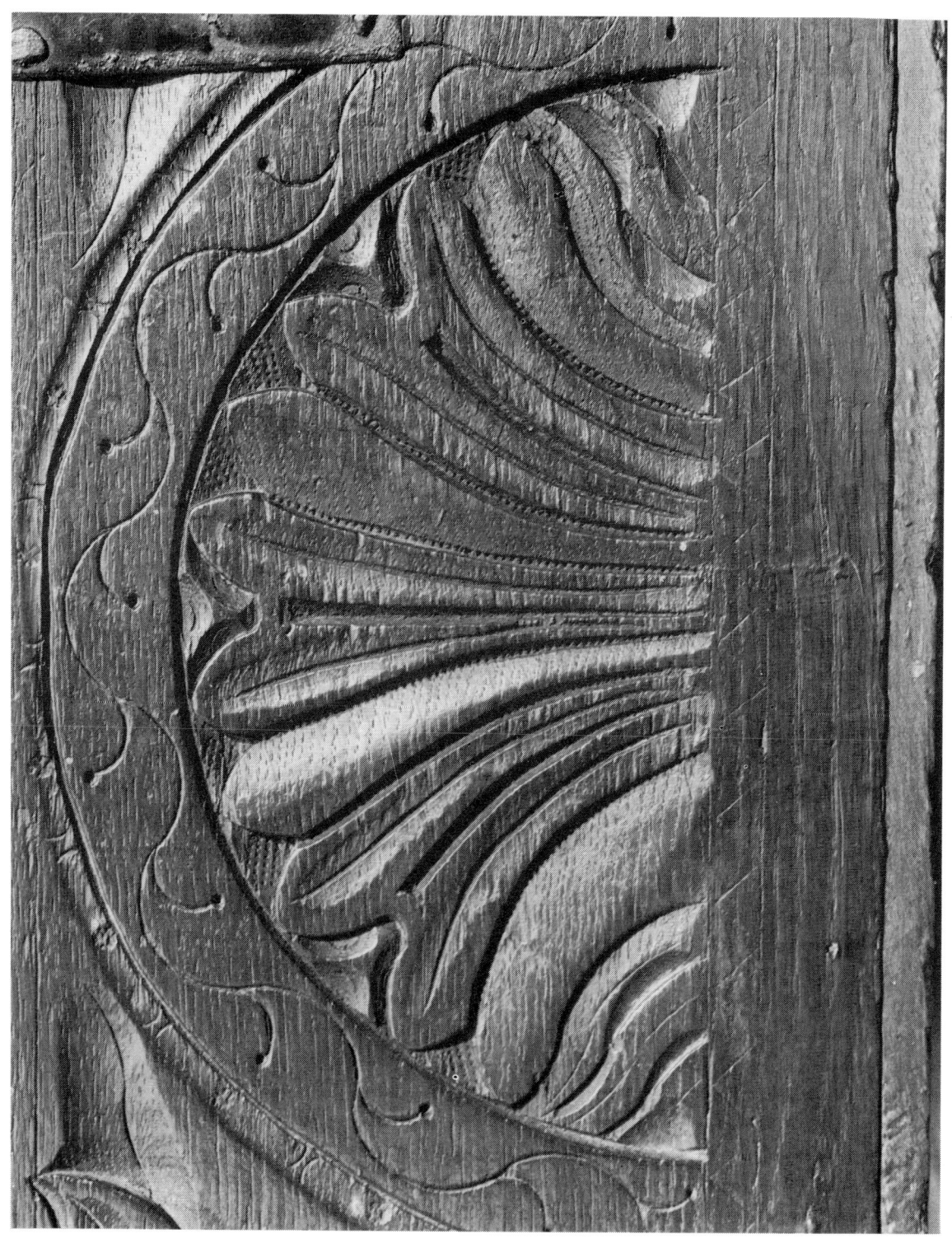

Detail, box no. 28.

61

The two boxes illustrated on the facing page are both obviously "country" in origin. Gone is the immaculate layout and precise carving which characterised their predecessors; gone too are the punch decoration and the scratch-block moulding.

Box no. 29 is neat geometrically but virtually the whole design was made with the "V" tool, when I first saw it I wondered whether someone had simply "roughed-out" the outlines prior to finishing the work with a couple of curved gouges, but no, the whole box was like this. (For convenience I illustrate the side panel only).

This lower box looks very like a copy, made by a fairly unskilled hand. The whole design has simply been drawn freehand upon the front panel and then taken out with a "V" tool. The maker also had a few curved gouges which he used to outline the foliar elements within the lunette bands, and to chip out the finger-nail shapes within the bands themselves. Despite its eccentric amateur appearance (or perhaps because of it) the box has a peculiar rural charm.

Box no. 29.   2nd/3rd quarter 17th century.

Box no. 30.   3rd quarter 17th century.

Box no. 31.   Dated 1694.

Here, right at the end of the 17th century, we have a motif that would have been quite at home a hundred years earlier. In fact, these "lunettes" are the Greek "palmette" design, possibly one of the most ancient decorative forms of classical antiquity. It was also an Egyptian motif and can be found in Mesopotamian ornament.

The design on this front panel however has little else in common with those on earlier boxes, it has been made using the late-century technique of simply drawing the outlines of the design on the flat surface of the board. To have a dated example is useful; the lunette bands are flat and carry hardly any decoration, in contrast to the earlier lunettes which are always embellished in some way, either by carving a channel along them, or by punchwork, or both. This "flat" feature can also be true of strapwork, (see box no.14 on page 33).

With this box we come to the end of the series exhibiting mainly classical motifs. Foreign influences on design have been evident on all the boxes and one wonders whether, if the Renaissance had taken some other form, the rich extravagance of the early carving could have been equalled. The various forms employed to decorate the desks and boxes lend themselves to deep carving, and this in turn gives a confident air of assurance to the work. The later carving lacks this confidence, as if the makers had lost direction, and it may be significant that most of the later carving seems to be "country" in origin. The new fashions in furniture which began to appear in the major centres of population in the 1660's were changes in the shape of furniture as much as in the decoration, and one can well imagine that the older more conservative craftsmen were slow to adapt. With the return of the second King Charles from abroad there was yet another burst of foreign influence which was to sweep away the Tudor and Jacobean forms for all time, together with all the strapwork, fluting, guilloche, and lunette decoration that had enriched the homes and furniture of the English middle-classes for generations.

TWO

Aids to the dating of early boxes.

Quite apart from the classical motifs employed by the carvers and the moulding sections used, other evidence can be gleaned from the boxes which helps to place them in a chronological order. Any information from the timber from which they were made is of obvious value, but an examination of locks and hinges can also yield useful facts.

The lining-papers which are occasionally found are of less value, but can be of enormous interest. Identifying the sources of these is an under-researched field.

Before going on to examine the decoration of late 17th and early 18th century desks and boxes it will be helpful to consider all the above.

An oak panel which formed the lid of a 17th century desk. The "medullary" rays show clearly on the surface of the timber. In better work, oak exhibiting this "figure" was always chosen by the box makers for the highly visible parts of the box, the lid and the front panel. Very often these rays stand "proud" of the surface due to the erosion of the softer wood around them, caused by having been polished for several centuries.

With one exception, all the boxes and desks illustrated in this book are of oak, which was the timber most commonly used in the making of domestic furniture in England for a very considerable period of time. Certainly the Anglo-Saxons used it, and it was not to become unfashionable until the end of the seventeenth century.

However, the second most available timber in England after the oak must have been the English Elm, which flourished on the warm deep loams of the lowlands, and must have been used for boxmaking particularly in rural areas. Elm is sometimes found in the back and bottom boards of boxes and coffers, usually bearing the irregular marks of the pit-saw, but it is rare to find it in early work.

Many of the early desks and boxes which we find today are made from timber that was cleft, and these are all of oak, which cleaves easily. Elm is a very strong material with an interlocking grain, and it is not possible to cleave it into boards and panels; it has to be sawn. This may well be one reason why if elm is found in a piece of furniture, it is likely to be late; much of the oak for cleaving came from coppice sources which by the middle of the 17th century were becoming worked out due to over production for industrial purposes. There appears to be more sawn oak in later boxes too, and I believe that this was because the timber fallers were forced to harvest the larger older oaks which they had avoided perhaps for centuries. Only the straight-grained butts of the larger trees would cleave, the knotty second lengths further up the tree would not. These second lengths moreover would not be as long as the butts, and material from them would have been more likely to have ended up in boxes, where the length of the panels rarely exceeds two feet.

The necessity to saw elm was one disadvantage, but a more serious one was its susceptibility to insect attack. If desks and boxes were made from elm back in the seventeenth century, very few would have survived until the present time, having simply been eaten by woodworm.

The durability of oak timber is legendary, particularly when cleft. In Herefordshire, right up to the 1940's it was being used in gate making, and cleft oak gates can still be found in places where the use of wider farm vehicles has not rendered them obsolete.

The act of cleaving a log of oak also has a value for the furniture maker in that it automatically reveals the beautiful "figure-mark" which is such a feature of the surface of the timber. This marking is caused by the "medullary" rays which radiate through the log like the spokes of a bicycle wheel (when viewed in cross-section). The timber splits naturally along the plane of these rays. Cleaving is a far quicker way of revealing the figure in oak than the alternative, which is "quarter" sawing; this involves firstly cutting the log into four segments, and then cutting boards from each segment, moving the baulk for each cut and taking smaller and smaller boards from each side of it. This is a lengthy process, and only gives strong rays on the surfaces of the very innermost boards.

Whilst it was available, the oak timber from coppiced areas was the most suitable for cleaving. It was fairly fast-grown, due to the fact that the trees were not overcrowded, and each was able to develop a full crown which in turn promoted growth. With a lack of competition from other oak, the trees tended to be short-boled with wide-spreading crowns, and the butts became clear of side-branches for perhaps no more than twenty feet.

The coppice system which produced timber of this nature had been in use in England since Roman times. Until the eighteenth century there was very little planting; woodland areas were simply fenced against stock and left to regenerate naturally, by seed or shoots from old stumps. Very early on it was discovered that these areas could be managed on a sustained-yield principle, in that if they were cut over regularly within a certain period of time they could be cut again. The interval between cuttings is a "rotation".

On a very local level, everything from these cuttings was used, from the bigger trees right down to the twigs and foliage. Also at an early date, it was discovered that if selected stems were left here and there within a coppice area at the end of each rotation, useful timber-sized trees would develop, and it was possible to take several crops of the understorey before the bigger trees were felled. There is evidence that rotations of coppice tended to increase in years as time went by; in the Middle Ages, a seven year interval between cuttings was normal, but when one gets into the 17th century rotations could be as long as thirty years.

This system which produces large timber as well as minor produce on the same area is known as "coppice-with-standards", and when it served only rural areas it proved sufficient to fulfill the needs of a largely peasant population. When in the sixteenth century the population increased and began to make demands on the woodland areas for timber for house-building, and charcoal for the manufacture of glass, bricks, tiles, and other fabricated essentials, the system broke down to such an extent that oak from coppice sources became in short supply.

In my previous book "Trees, Chests and Boxes of the Sixteenth and Seventeenth Centuries" I examined a sequence of boxes from about 1550 to 1750 with a view to assessing whether they were made from oak from coppice sources, or from "wood-pasture", which produced longer lengths of timber not unlike that which we produce today under our "high-forest" system.

By examining the "end-grain" of the boards from which boxes were made it is possible to deduce from the annual rings whether the trees from which they were made grew at a steady rate, or had periods of slower growth; these show up as sequences of narrower annual rings.

The cross-section of a board from box no. 27, (page 58) illustrated on the facing page shows three distinct sequences of narrower rings.

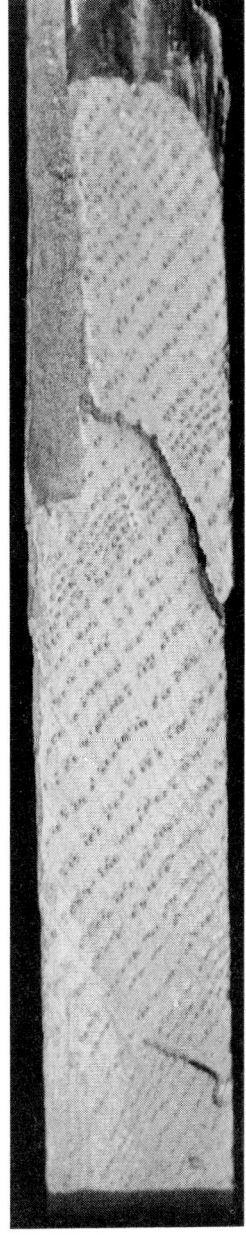

Five annual rings exhibiting slower growth.

Eight rings showing faster growth.

Five annual rings exhibiting slower growth.

Thirteen rings showing faster growth.

Five annual rings exhibiting slower growth.

The curvature of the annual rings indicates that the growth in this board is from the bottom of the photograph to the top. The lower "rotation" is eighteen years; the higher is thirteen.

69

In my sample, eighteen of the boxes dated from about 1630-1640, and of these, fourteen showed no evidence of a coppice origin for the oak from which they were made. In statistical terms the sample is too small to justify a firm conclusion, but we do know that a serious shortage of coppice-grown timber had developed by the 1630's; supplies of charcoal had diminished to such an extent that even the smelting of iron to make cannon had to be curtailed. The price of charcoal more than doubled between 1630 and 1670.

So it may be correct to assume that if one finds evidence of a coppice source for the oak in a box, it is likely to be early.

To return to cleaving: my sample of the same boxes also sought to differentiate between the presence of sawn or cleft timber in them. Here again, of the eighteen later boxes, seventeen were made from oak which had been sawn. Of the early boxes, half of them contained cleft material, three being made entirely from cleft oak.

It is not too difficult to differentiate between sawn boards and cleft. The act of cleaving always leaves a ragged surface on the panel, where the grain of the wood has been ripped apart. The early makers were careful to finish the upper visible surface of boards by planing or scraping which left a smooth surface, but they were not so conscientious over the lower surfaces which were out of sight, and it is here that you can find evidence of the cleaving process. Also, the thickness of cleft panels tends to vary, even within one box. Sawn material on the other hand is always of the same thickness.

The lids of early desks and boxes were very often thin, 1/4" being almost a "standard" thickness, and the fact that they have not warped and split testifies to the seasoning of the timber prior to use. One occasionally finds split bottom boards which would seem to indicate that from time to time, if pushed, the maker would use some younger stock to complete a box. Timber always shrinks across the grain, and an unseasoned piece, once nailed, would almost certainly split as it dried out.

One other fact that emerged in my sample of boxes was that half of the early ones contained timber with a very slow growth rate, varying between twenty and thirty-two rings to the inch. These boards were very often chosen for the front panels, I presume by the carvers, who would seek the finer grain to give a crisp edge to their work.

It is difficult to imagine that rates of growth as slow as these were produced by English trees. The average rate of growth of oak in this country is about eight annual rings to the inch, and it seems likely that this slower-grown oak was imported. We know that Polish oak had been imported into this country from medieval times. The sample illustrated below is from the back panel of box no. 3, (page 16), and is full-size.

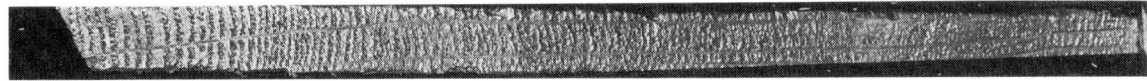

Widely-spaced "standard" oaks in a Herefordshire coppice. The understorey of younger smaller material has been felled and removed. The trees have wide crowns, and short butts; these however are of the highest quality, being free of knots, and would cleave well. (The area is being enriched by new planting in the tubes).

Wood-pasture oak in the Forest of Dean, Gloucestershire. The labour involved in felling and removing trees of this size was considerable; very often they were converted into planks on site, hence the number of "Sawpit" woods still with us today. Before powered sawmills existed it was cheaper to take the saw to the wood. Trees of this size were avoided if "standard" coppice-grown oaks were available.

There are local differences in the rate of growth of oak which show up in the end-grain of the boards from which desks and boxes were made. Trees growing on good soil in warm, sheltered locations in the Westcountry grow relatively quickly; in the drier colder climate of East Anglia they do not.

Trees growing in high rainfall areas such as South Devon can put on as much as 1/4" of timber in a year; a similar tree in Norfolk may only achieve 1/10". As home-grown timber was almost all bought locally in the period of our study it may be possible, by matching rates of growth to styles of carving, to confirm these as belonging to a particular area.

The extremely fast rate of growth shown (full size) below was put on by the tree which produced the timber from which the back board of box no. 1 (page 9) was made. I do not believe it possible for a tree from the east of the country to produce this rate of growth. On the box, the carving of the opposing lunettes beneath the lock plate is thought to be a Gloucestershire feature; if this is so, the rate of growth is consistent with that of a tree growing on a fertile site, perhaps in the Severn Valley.

The fact that the annual rings in this board are at an angle to the surface tells us immediately that the board was produced by sawing. Cleft oak always has these at right angles, which is another way of confirming whether a panel was made from cleft timber.

The use of ring-widths to establish a locality for the timber needs to be treated with caution; trees on opposite slopes of the same valley may show different rates of growth; this will also vary with altitude and exposure. However, in general terms there is sense in the hypothesis, and it is well worth while examining the end-grain of boxes for any information which may lie there.

Dendrochronology may not be of much use to the furniture historian, particularly where coppice-grown timber is concerned, as the variations in ring widths caused by the cutting rotation will mask the evidence of differences due to climate. The timber in the imported boards however may well yield useful information as this was produced in European forests; detailed dendrochronologies exist which may enable dates to be established for the sequences of rings, which in turn will give us some indication of the time when the timber was made into a box.

To establish this precisely will still not be possible, due to the unknown factor of just what width of sapwood was discarded by the maker; this can be as much as two inches in width, representing perhaps forty years of growth.

Another factor may help to establish a region of growth, if not a locality. This lies in the distribution of the two species of oak resident in England.

The "sessile" oak is more often found in the north and west; it is thought to be the commonest oak in Wales. It succeeds on drier sites and is less affected by frost and insect attack. It is less prone to produce epicormic buds and thus produces longer lengths of knot-free timber. Evelyn thought it best for timber as it grew "more upright". "Pedunculate" oak however, is a tree of the deep, fresh, fertile soils, including clays; a lowland tree. It is the oak which produced the Anglo-Saxon pig-pastures, and, due to its branching habit, the ship-building timbers of the Elizabethan navy. Identification in the field is easy; apart from its upright habit, the sessile acorn sits on the twig, the pedunculate is stalked. Within the timber however, one needs a microscope to see the oval section of the sessile cell, compared to the round of the pedunculate. Again, however, even if we identify the species it is still not possible to accurately identify the locality in which it grew; the distribution of pedunculate oak is wide-spread, and was the coppice tree of the old woodlands.

Although we can gain a great deal of information from the timber from which desks and boxes were made, many of the conclusions at the present time are speculative. However, it is difficult to see what Forestry system would produce the pattern of annual rings illustrated on page 69 if it was not coppice. One finds these sequences time and again in early desks and boxs.

The stalked acorns of pedunculate oak, carved on the side panel of box no. 33, (page 95).

HASPS, LOCKS AND LOCKPLATES.

Unless you are incredibly lucky you will not find lockplates of the quality illustrated on the previous page, this comes from a mid–16th century six-plank chest. Most of the locks encountered on the smaller desks and boxes are fairly mundane, and unhappily for the researcher, show no demonstrable thread of evolution for perhaps two hundred years.

It is obvious from a close study of many boxes that a considerable number of replacements were fitted over the years; to complicate matters even further in many cases only the hasp was replaced, on some examples perhaps twice.

To attempt to sort out a chronological sequence of locks is therefore virtually impossible. Where locks can assist us is when one was replaced at a date which can be reasonably established; it is reasonable to assume that the box itself was made well before that time.

It is difficult to establish just why hasps wore out faster then the original locks. The mechanism is simple, the hasp which fits into the slot of the lock is attached to a strap which is nailed to the lid of the box, usually inside. It hinges around a soft iron pin which is riveted at each end to stop it dropping out. If a key were lost, the hasp is the obvious candidate for attack, but in many cases all that would be necessary to open the box would be to drive out the pin. This would of course still leave the hasp secured to the lock. Following this through would seem to indicate that sheer wear was the most likely cause of replacement.

Forcible entry to the box would result in the strap being detached from the lid, and to replace the original strap would entail using the same nail holes. In these circumstances it would be more secure to fit another hasp and strap where the nails would go into the wood in different locations.

Fortunately it is not difficult to determine where this has happened, the original nail holes (and sometimes the rusted stumps of the nails themselves) will remain in the wood. This also holds good for replaced hinges.

In the illustration opposite, the lock is the original, but the hasp and strap are replacements. The first clue is in the wide slot cut in the edge of the lid, this is much wider than the replacement requires.

Secondly, the strap is shorter than its predecessor by perhaps an inch and a half; the stump of the old nail remains in the wood, and has stained the oak surrounding it. This staining is characteristic of oak; the iron of the nails reacts with the tannin and other chemicals in the timber to produce unmistakable evidence of a foreign body.

There are no suspicious nail holes around the lock plate, although these could be hidden beneath it if the original lock was smaller. If this was so, it is usually possible to find staining inside the board to which the lock is secured.

A nail has disappeared from the top right hand corner of the plate, but this registers exactly with the empty hole behind.

Using the above it is possible to determine the "lock history" of any box or desk.

Lock and hasp on box no. 11 (page 29).

The two illustrations on this page are both of locks and hasps which were fitted to their boxes at the time of manufacture. There are absolutely no signs anywhere of odd nail holes or patches of stained wood, inside or outside the boxes.

The box on the right was made perhaps in 1600, and carries the standard "clicket" lock which had already been in service during the previous century. It is from box. no. 17, (page 38).

The lock on the right is from box no. 15 (page 34), which cannot have been made much earlier than the last decade of the seventeenth century; it could equally well have been made in the 1720's given its country provenance. I have seen an identical lock on a box dated 1795.

The external similarity of the two locks encapsulates the difficulty of dating by lock style alone. Internally, however, they must differ; the keys are not interchangeable.

The lock on the right is from box no. 6 (page 21) which is almost certainly not the original, although the evidence lies only in slight discolouration of the oak to the left of the plate. The hasp, on the other hand, is certainly the second, and as decoratively it seems to match the lock plate it seems likely that both were replaced in perhaps the 1650's. At around the mid-century some elaboration seems to appear in the design of lock plates which occurs here and there on smaller items of furniture. The elaborate plate overleaf is from a box of about 1650.

The lock on the right is from box. no. 33 (page 95) and again both it and the hasp are original. Although the carved motifs on the box can be found early in the century, this is a "country" piece and I would put it somewhere in the last quarter of the century. If I am right this could serve as a useful benchmark for dating other boxes with similar hasps fitted at the time of manufacture.

Detail, box no. 37, (page 101).

The lock on the right is from box no. 39 (page 102), which has within it some dated lining paper from 1653. Both the lock and the hasp however are replacements, which is unfortunate, as had they been originals the correlation with the paper might have been significant. We could estimate, however that this shape of hasp might be from the early 1800's.

The lock on the right is the original, and was fixed to the box in 1694, which is the date carved on the front panel of the box. Compared with the carving, which is excellent, the lock is crude, but in shape is not unlike that on the previous page.

At this date, much of the "new" furniture bore decorative escutcheons, usually in brass. Was this an attempt by a maker to keep up with the competition?

The hasp is a modern copy.

The lock on the right is clearly not the original; beneath it is a plugged hole which received the back of the first lock. Furthermore the keyhole bears a shield–shaped brass escutcheon. It appears to be only the second fitted to the box, in which case the first lasted quite a long time. The box itself (no.20 page 44) is very early Jacobean; the present lock is Victorian. The hasp is contemporary with the new lock.

This particular shape of hasp is very typical of later additions to boxes and desks; the brass feature on the lock is also characteristic, one can occasionally find a lock where this has fallen off, but this reveals riveting slots around the keyhole.

This lock is from a box dated 1711, (box no.16, page 35). Both the hasp and the lock are Victorian however; indeed the word "patent" is inscribed on the keyhole cover. Many of these later locks are partially chamfered along the edges of the plate; both this one and the previous example have this feature.

82

From all this it is possible to put together a "check-list" of features to examine when confronted with the lock on an unfamiliar desk or table box.

One: is the lock that which was originally fitted to the box? Check for alterations to the wood surrounding it; are there unexplained gaps below or above the lock plate? Check also inside the box; where the front panel is thin the back of the lock is sometimes visible. Are there signs of a larger space for a lock of different shape?

Two: look carefully at the wood surrounding the lock plate for signs of old nails; pay attention to any discolouration, especially if of a regular shape, which might have been caused by being beneath a larger lock plate for some earlier period. Again, look inside the box for stained places which might betray the location of earlier nails which secured a smaller lock plate.

Three: Examine the hasp. Does it easily fit the slot in the lock?

Four: Check the hasp strap. Sometimes these fit neatly into a recess of the same shape chiselled into the lower surface of the lid. Does it fit exactly? If it is not recessed, are there any old nails or nail holes away from the present location of the strap? With straps fitted beneath the lid (the majority) it is sometimes possible to detect the presence of rogue nails on the top surface, by stained patches in the oak.

Five: check that any recesses cut in the top of the front panel to accommodate the strap actually match the present position of it.

Six: if there is any evidence of replacement, try and assess whether this is of both the hasp and the lock.

Seven: how many replacements have there been?

Eight: now look at the carving and try and tie it in with the evidence from the lock.

Personally, I do not mind if a box does not have the original lock and hasp, indeed one is most suspicious if old furniture has not seen some repair. To unravel part of the history of a box can be most gratifying, and in most of us there is a small detective trying to get out. There is a distinct pleasure in satisfying one's curiosity.

Facing page: 17th century wood-block-printed lining paper. From Charter-boxes in Corpus Christi College. The boxes were made in 1627, and the paper is probably contemporary. Although the flowers are English, there is a decidedly "arabesque" feel to the design, mainly due to the "S" curves of the plant-stems.

(Courtesy Ashmolean Museum, Oxford).

More "S" scrolls, these are from the initial letters at the beginning of text from a Psalter published in London in 1615.

LINING–PAPERS

85

The practice of putting paper into table boxes is probably as old as the boxes themselves; particularly when they were used to protect articles of clothing, such as ruffs, from dust. That so few boxes still contain lining-paper is probably due to people just placing this into the boxes, instead of making a job of it and using glue or paste. Even when pasted in, many papers must simply have disintegrated due to the passing of time.

To find an interesting old paper in a box can be exciting, but unfortunately is of little help in accurately dating a box. There is no "starting-date" within our period before which no printed papers were produced.

The practice of printing designs on paper probably dates from the end of the fifteenth century, using engraved wooden blocks; the earliest secular use of papers so produced was to cover books in place of leather, which was expensive. (There are those who consider the paper-back book a *modern* invention).

The same papers were also used to line chests and boxes and as "wall-paper" to decorate rooms in place of the more expensive hangings. All the early papers were printed in black, upon a coarse-textured paper, but the paper itself was occasionally coloured.

Many Deed and Charter boxes were lined with paper; examples exist from the late 16th century onwards.

A distinction can be made between lining-paper and wallpaper in that the sheets of the latter carry incomplete designs until they are matched with another sheet. Tudor and Jacobean papers were printed with a paper size of between 14" and 16" wide and 11" long; rolled wall paper was a late 17th century innovation. If one finds a lining paper of the correct size with a complete design, it follows that this was not wallpaper. Using the dimensions, it may also be possible to say whether the paper is early or late.

The paper illustrated on the facing page lines the bottom of box no. 36, (page 99). The larger block printing with the dolphins is approximately 16" by 14", above it is part of a sheet of paper depicting the four seasons; the margin of this is pasted on top of the larger sheet. The four seasons print is part of the Arms of the Haberdashers' Company, which was incorporated in 1689. This puts a neat date to the paper, but it does not date the box. A provenance for the design on the larger sheet has so far escaped me, but it would help if it could be shown to be some thirty years earlier, which would equate with when I think the box may have been made.

One problem with these patterns is they appear all over the country; there does not appear to be any regional character that can be recognised. Regional carving styles can be identified for some parts of the country, but even when these are discovered in conjunction with lining papers there does not seem to be any useful correlation.

Another problem is that very few of these early papers have survived, either as lining paper or wallpaper; the latter rarely survives frequent redecoration of houses.

The difficulty in dating by lining paper is well illustrated by the above. The badges are those of the Prince of Wales; in this case identified by the initials H.P. as Henry Frederick, (1594-1612). He was created Prince of Wales in 1610, just two years before he died. The papers were found in a box dated 1635. How does one explain the twenty-five year interval?

I cannot believe that some printer solemnly ran off these papers for years after Henry Frederick died; nor is it likely that at the time of the death he printed such a quantity that it took twenty-five years to destock.

Yet if the box in which they were found had not been dated, and had it exhibited early features in the carving or the moulding, it is by no means impossible that someone might see the papers as proof that the box had been made between 1610 and 1612.

The above is a paper "dated" by implication. On the facing page is some lining paper which actually carries a date. This lives in box no. 39, page 102. The whole of the bottom of this box is lined with uncut sheets of pages from a book called "Astrological Judgements", giving predictions for each month in 1653, ending in December of that year. These have been overprinted before being pasted into the box with a blackwork pattern of flowers, leaves, and scrolled lines.

Bearing in mind what happened to the H.P. paper above, is one safe in assuming that the box was made in the 1650's ? The box itself has a fleur-de-lis motif which is very difficult to date as it was popular for a considerable period of time. I have seen the identical carving on a dated chair of 1620. Perhaps the best we can hope for is to assume that 1653 is the latest date for the box. This does not seem unreasonable, and the element of uncertainty gives a great deal of interest to the box. Did it originally belong to a printer?

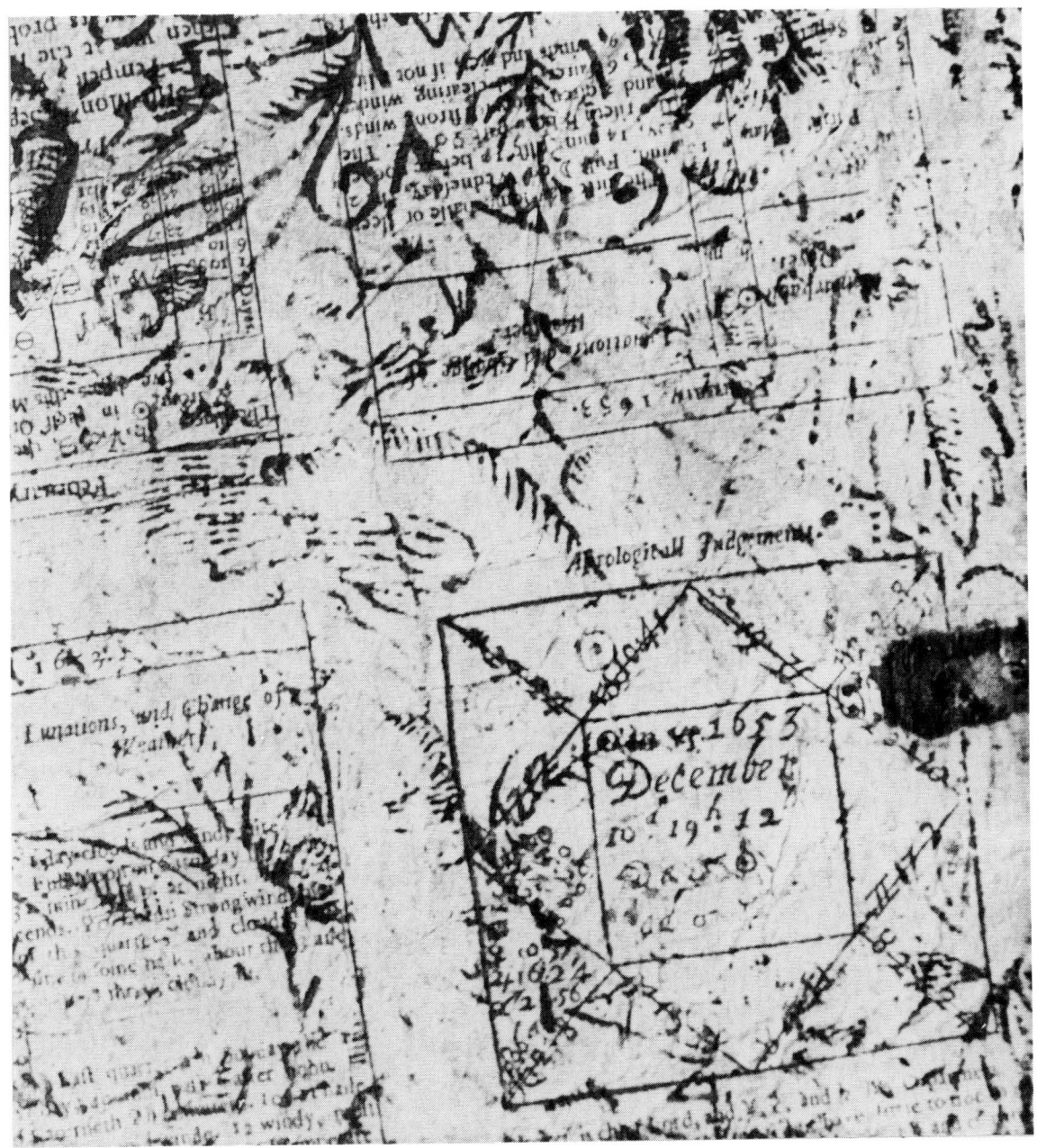

I suppose that today the most common lining-paper used for chests of drawers and wardrobes is newspaper; it is always to hand and has already been paid for. The Victorians were great at economy and did the same. If one finds a dated local newspaper from a hundred years ago in a box, does this have any significance?

I believe that it may have; families then were not so mobile as they are today, and the middle classes on the whole probably stayed in the towns and villages where they had been born.

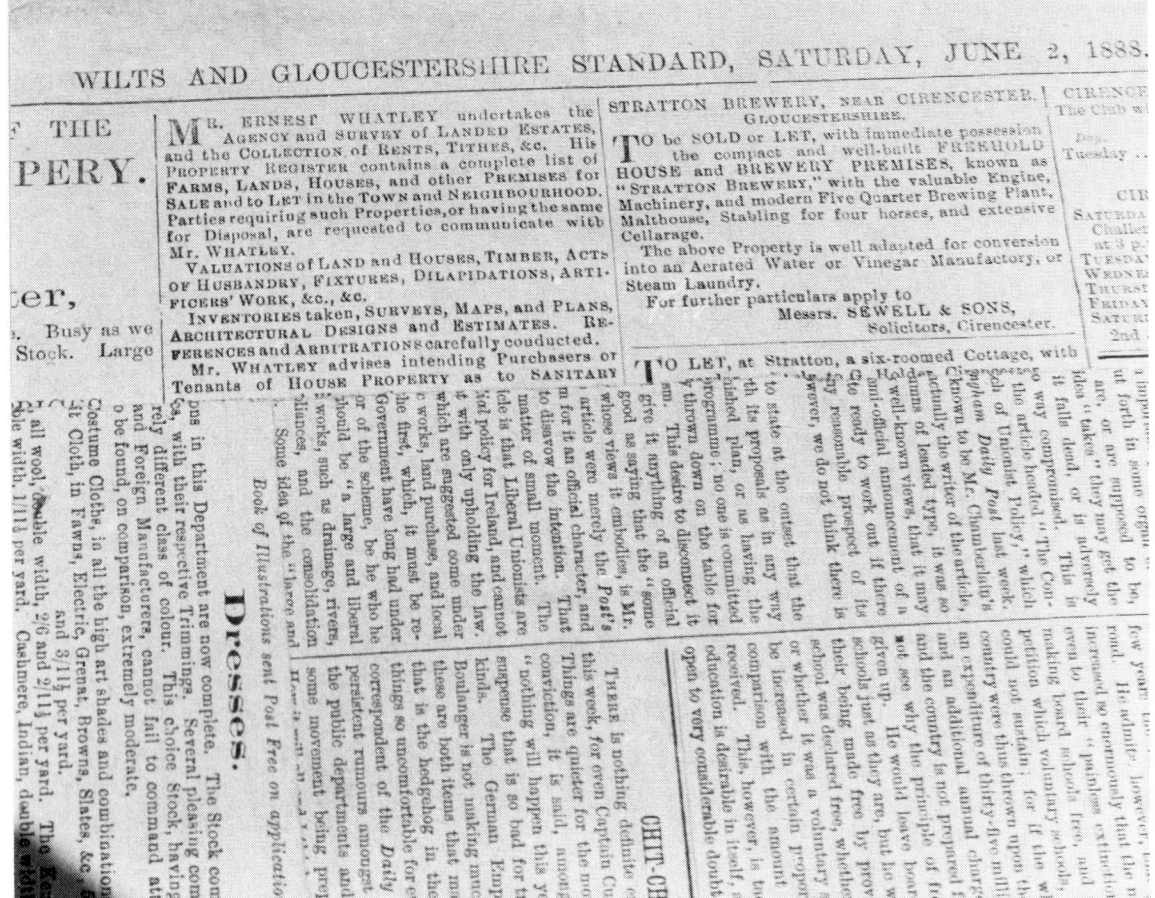

The above newspaper is glued to the bottom of box no. 1 on page 9. The box was in Tetbury, Gloucestershire, in 1990, and there is just a chance that it had always lived in that area. The great shuffling of antique furniture that has taken place in the past hundred years has made the tracing of regional styles of carving very difficult. Here, if we assume that the box had not travelled far from its original home, we may have confirmation of a regional style from the newspaper. This would bear out the information from the annual rings in the timber in this box, already discussed.

It follows from all of the above, that no detail in a box, whether in the timber, the locks, or any lining paper is unimportant if we are to attempt to discover its history. Taken in conjunction when several factors come together they can by association yield significant information, which if one is lucky may place a box at a certain period or location. Oddly enough, the earlier boxes may be easier to place than the later; when we come to the desks and boxes of the late 17th century a degree of uniformity appears which adds to their anonymity.

90

THREE.

Non-classical motifs, and later desks and boxes.

Changes in design of course did not happen overnight, and there was a considerable period of overlapping.

The factors which led to changes in fashion were once again foreign. If we look at the changing fashions in the clothes which people wore over the whole period of our study, we can identify the most important periods of change.

In the 16th century a stability in society emerged at last from the depredations of the various plagues, and the Civil War. Fashions in clothing were dominated by the Court and the nobility. In the later years of Henry VIII these were very much influenced by the Spanish predeliction for drab colours and these were adopted by the English, particularly after the marriage of Mary Tudor to the Spanish King. The effect of these rigid fashions was to present an arrogance in posture which was copied with enthusiasm by the new members of the aristocracy who had very often emerged from the old middle-classes.

As the century drew on however, the need for extravagant display in dress increased; costume was used to denote class distinction. "Power-dressing" was in.

At around 1570, the ruff had developed from the collar-bands of mens' shirts and was a separate article, increasing in magnificence as the century progressed. Similarly, up to about that date headgear was low and flat, changing into the hat thereafter.

These trends in fashion were reflected on the furniture, which became encrusted in carved Renaissance motifs.

The 17th century saw a decline in the Spanish influence; up to about 1630 the fashions in the clothing of both sexes were very little different from those of the late Elizabethan period. When Charles Ist married his French queen she brought with her her own courtiers and they in turn began to influence English fashion. At about 1635, shoes which had hitherto been rounded became square. By the 1630's the ruff had become unfashionable. Gloves became fringed by lace. From quite early in the century the Cavaliers had copied the French fashions, unlike their cousins the Roundheads who were more influenced by the Dutch (who cropped their hair; hence the name).

During the Civil War period there was little change in costume, but upon the return of the second King Charles from abroad, French fashions once again reigned supreme and extravagance again became the norm. In 1666 the entire Court, by direction of the King, changed fashion, adopting an extraordinary mode which Evelyn described as "Persian". Oddly, this exotic form of dress was to create the forerunners of the modern waistcoat and overcoat.

By the 1690's, the skill of the tailors had improved to the point where the cut of a man's clothing could denote his social status.

Thus in clothing fashions we have three "benchmark" periods, the late Elizabethan years, the 1630's, and the years following the return of Charles II.

I believe that the same periods of change can be detected in the carving of the desks and boxes; certainly many of the later examples have a very fussy "French" look to them. One can find a Dutch influence, too, after about the 1660's, when tulips appear as a decoration; many lantern clocks of this period carry them engraved upon their dials.

There are other influences upon design too, quite apart from the fashion in clothing. The Delftware chargers of about 1680 carried formal arrangements of tulips and carnations; these were collected and displayed with enthusiasm. In 1660, the East India Company began to import lacquer wares in quantity, and very soon these were being imitated, usually badly.

By the end of the 17th century, the whole country was more prosperous than it had ever been before and many more people were able to furnish their homes with goods that hitherto would have been regarded as luxuries. It has been calculated that at that time, people were ten to twenty times richer in furniture and other domestic items than their grandfathers had been. Gregory King calculated in 1688 that their were 36,000 families in England with an income of over £200 per annum. It can be no coincidence that 1694 saw the founding of the Bank of England, nor that today we find many more late boxes than early ones.

All this is a long way in philosophical terms from the 16th century, when men still believed in the classical doctrine of living in harmony with nature. From quite early in the 17th century, scientists were beginning to invent for themselves a wholly mathematical universe; Newton and his contemporaries gave this authority in a process that has continued to the present day.

This change in philosophy shows up upon the desks and boxes of the late 17th century; where naturalistic motifs are employed, they tend to be restricted and purely representational. Instead of the Renaissance motifs which flow across the whole of the front we have floral patterns confined to a panel on each side of the lock plates. Geometry replaces the ancient motifs from the Mediterranean.

Before moving on to the designs of the late century we must consider a group of boxes which cover the transitional period from about 1630 to mid-century; these show an interesting mix of the older methods of decoration left over from the preceding period, but using motifs which are no longer wholly classical.

Box no. 32.   2nd quarter 17th century.

The upper decoration on this box has a strong regional connection, other items bearing the "buttercup" motif have been shown to come from the Wiltshire/Dorset area.

The buttercup, and the thistle on the lower half of the front panel, are not in any way classical in origin; the thistle could even have been deliberately employed as a Royal emblem of the Stuart family.

The timber content of the box hints at an early date. One of the side panels was made from cleft oak, whereas the rest of the panels are of sawn timber. The regional origin may be confirmed by the fastest rate of growth of the oak, at seven rings to the inch, which is in the bottom boards, these also carry evidence of a coppice source, with a rotation of 25 years. This long rotational period equates with some of those found in other boxes of the 1630's.

There is no shallow carving on the front panel, the petals of the buttercups have been excavated to a considerable depth. So much so, in fact, that the carver ran the risk of breaking through the material completely; the bottoms of the petals are paper-thin.

The front panel has been carved into a piece of oak which may well have been imported; the annual rings are tightly packed together at about 32 to the inch.

There is a lot of punchwork decoration within the elements of the design; the depths of the petals carry similar punchwork marks as those found within the guilloche banding on box no. 8 on page 25. There is liberal use of a small round punch on the raised parts of the surface.

The lid of the box is unfortunately a modern replacement and we thus have no information concerning the original moulding, but it would probably have been made with a scratch-block.

There is no sign that there ever was any moulding along the bottom of the box.

Although decidedly "rural" in appearance, the carving is of quite a high standard and I think it would be unfair to give it a country designation. At some time it has been sadly neglected, which accounts for the rather rough appearance of the timber.

Box no. 33.   Mid/late 17th century.

This is another "buttercup" box, slightly larger than box no. 32 and perhaps thirty years later in date. The outlines of the elements of the design have simply been drawn with a "V" gouge and there has been no attempt by the carver to excavate the petals. There is a minimum of punchwork decoration.

It has been constructed entirely of sawn timber; one of the side panels exhibits a very fast rate of growth, at perhaps five annual rings to the inch. These side panels bear the carved acorns and oak leaves illustrated on page 74. In view of this, and the association of the front panel motif with Dorset and Wiltshire I feel safe in claiming a Westcountry origin for the desk.

The little fence at the top was added later, presumably to stop things from dropping down the back.

Originally the desk had a deep bottom moulding; one can see the remains of nails at the bottom of the front panel, on each side of the lock plate. From the position of the nails one can determine that the bottom of the lock plate must have sitting in a slot cut in the moulding. With this depth, the moulding must have been very like that on the next box, no. 33 overleaf. It is a great pity that it is missing, if still in place it would have enhanced the appearance of the desk considerably.

There is no top moulding, the edges of the lid have simply been rounded-off.

A geometric style has crept into the carving, which is quite skilfully done, all the curves of the petals have either been marked out with a compass or drawn around a template. On this front panel there are no signs of the point of a compass having been pressed into the wood; one can probably assume the latter. This in turn implies that there might have been an early form of mass-production; one would hardly make a template for a one-off job.

Box no. 34.   2nd quarter 17th century.

The carving of the panels on this box is of the highest quality; the carver has achieved a sculptural effect by very slightly chamfering the surfaces of the leaf-shapes prior to decorating them with punchwork. There is an assured crispness to the work, which must be that of a highly-skilled craftsman.

The box has many claims to be early. The lid bears the shallow half-round section illustrated on page 20, made with a scratch-block. The deep bottom moulding is characteristic of better work of the early 17th century. The lid is made from cleft oak and displays medullary rays, although all the other panels in the box are sawn. In addition the lid has strong evidence of a coppice origin, showing a sequence of twenty-two years of rapid growth between two periods of slower. In part of this sequence of rapid growth the tree achieved five annual rings to the inch, indicating a warm moist location.

The carved front panel came from an entirely different tree, which grew at an average rate of only sixteen rings to the inch, hinting again at a deliberate choice on the part of the carver.

The punchwork decoration is profuse, extending even to the marking out of the notches on each side of the front panel with a very small pointed tool.

I have been told that the leaf-like motif represents a "wheat-ear" pattern and may very well originate in Devon. The rate of growth of the timber certainly seems to bear this out.

The lock and hasp on this box are 18th century replacements, but the hinges are original. These are of the "fishtail" variety which were in common use for perhaps two hundred years and thus are no guide to dating.

As well as the half-round moulding, the lid of this box also bears a punched decoration in the margins composed of flattened "S" shapes, perhaps 3/4" inch wide. These marginal lid decorations appear to come in at about this time, and persist for perhaps the next half-century. (See page 140).

This is a useful example of a "transitional" box, bearing as it does a firmly English design.

96

97

Box no. 35.   Late 17th century.

At first sight this desk appears to have gone back to lunette decoration, but here we have the flat banding already noted (see box. no. 31, page 64). In outline, the design within the bands is the same as that on the previous box, and I feel safe in giving it a late date.

Here too we encounter our first tulips, which in turn indicate a post–Restoration date. The side panels carry a similar decoration, which is nearly always the sign of a "quality" box.

When one looks at the timber in the box there is a remarkable similarity to that in box no. 34 (page 96). The maximum growth rate in each box is five rings to the inch. In addition, the punch decoration is similar in that each uses a punch in the shape of a small four-bladed propeller (or flower).

The decoration of the side panels in the earlier box carried a design very similar to that on the above desk, as if the earlier box was a forerunner of the later. The deep bottom moulding on each box has the same section.

In view of all this, it is not impossible that they came from the same workshop.

The internal lock is interesting; the present one is Victorian. There is no evidence of an earlier lock in the woodwork, and there are no nail or screw holes behind the front panel which might have secured an earlier lock. One has to consider that originally the desk had no lock at all.

The lid of this desk shelves steeply and hinges directly from the top of the backboard. The customary shelf at the back of the lid is absent, which I believe may be a later feature. If rooms were better lighted there would be no need for a shelf on top of the desk on which to put a candle. (When the lid on these desks was open, the contents were in shadow. One finds many desks and boxes with charred patches on the inside of the lid caused by placing candles inside the box).

98

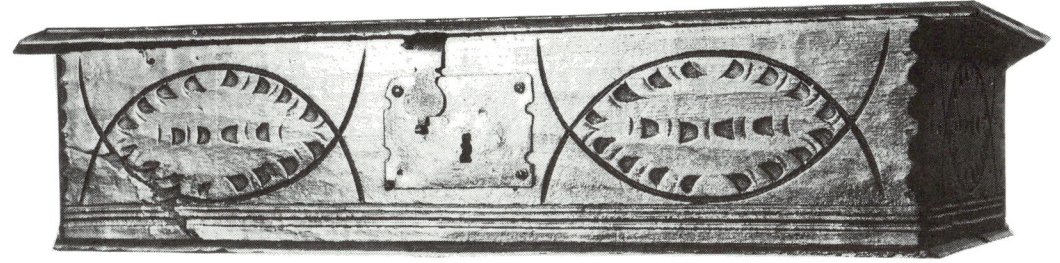

Box no. 36.   Mid 17th century.

Here we have another front panel where the design element is based on the ovoid; in this case achieved very simply by striking two intersecting arcs with a "V" tool and chipping out the internal decoration with a ½" diameter gouge. The same design is used on the side panels but with the addition of punch decoration in the shape of a small star.

All the timber in the box was sawn, and shows a maximum growth rate of six annual rings to the inch.

The austerity of the design may put the date of this box at pre-Restoration; a similar form of decoration is used on other furniture in the Salisbury, Wiltshire area. The motif is very like that of the early Christian "fish" symbol.

The channel moulding running along the lower part of the front and side panels is scratch-block work.

The box is lined with paper bearing the Arms of the Habedashers' Company, and if this was done when the box was new, then my suggested dating is wrong. (See pp 86 and 87). One objection to a later date is the depth of the box, which is only 6½", later boxes seem to average 10". On the other hand the nails are quite small and neat.

The presence of paper from the Haberdasher's Company is interesting. Was this originally wrapped around a pair of gloves or something similar which were then kept in the box? In which case we may have another clue to dating. Chests of drawers must have driven the smaller boxes out of the bedroom and into the hall at some time, where they would be very useful for containing the more "outdoor" articles of clothing, such as gloves and bonnets. The depth of the later boxes may have been increased to contain the latter.

The shallow depth of this box, together with the "fish" symbol, may also mean that it is perhaps the only "Bible" box in my sample, in that it could easily contain the Bibles of the period.

Box no. 36, opposite, is another "transitional", the carver has created a hybrid between the classical guilloche banding and the rather stylised leaves of the English oak. Here again, the timber may help to locate the original source of the box.

The lid and the bottom boards are both of cleft oak; the lid is one single board 16" in width which displays the most vivid figure-marking across the whole width of the surface. This gives us an opportunity to arrive at an average rate of growth, the tree took 86 years to make the lid, or an average of 5.4 rings to the inch. Again, we may be looking to the Westcountry for a provenance, and this is reinforced by the design of the quatrefoils on the panel; this particular motif was very popular in that region during the 17th century.

In addition to the above, the lid has scratch-block moulding around the edges, this is in conjunction with a marginal chevron decoration similar to that on box. 33, (page 95).

Within the box is a lining-paper which gives a list of the Fairs and Markets for Hereford and the border counties for 1824, and a window-tax schedule showing six windows.

I reproduce the lower illustration by kind permission of the Welsh Folk Museum at St. Fagans, near Cardiff. The quatrefoil design on this box is very similar to that on box no. 36 with its stylised guilloche banding, and I believe that it is a pattern that must have crossed the Bristol Channel to its present home. There is no reason why the box itself may not have been made in Wales; it may even have been a copy.

The date seems to have been added later, the figures occupy the spaces in a haphazard manner which the carver of the quatrefoils would not have done. In view of this, one can probably safely put the box into the very late 17th or early 18th centuries.

Box no. 37. Mid 17th century.

Box no. 38. dated 1721.

Box no. 39.   Mid 17th century.

This is a very difficult box to date, the fleur-de-lis motif has a long history and was very popular during the 17th century. In Christian iconography it is symbolic of the Virgin Mary, and as an heraldic device there is a French connection in that it appeared on the Royal Arms of France. I have seen it on a dated chair of 1620.

Within the timber from which the box was made there are no clues whatsoever as to whether it is early or late.

Within the box however there is an interesting lining-paper carrying the date 1653; see pages 88 and 89.

The original deep bottom moulding is missing, but there is evidence that it came almost up to the level of the carved panel. The patina on this lower section of the box however is virtually the same as that on the higher, and the moulding must have fallen off very early on.

The box illustrated on the facing page is geometrically similar in layout, which is compass-work. The holes of the centres from which the arcs were struck are still present in the wood of the front panel.

This is a very striking design for a front panel, and owes nothing to classicism. The lunette-like banding has been hollowed out with a gouge and there is some punchwork within the bands which hints at an earlier date, but the treatment of the flat surface of the wood, where a "V" tool has simply been jabbed into the surface to simulate perhaps the stalks of the wheatsheaves points more to the end of the century.

Could the scooped-out elipses within the wheatsheaf motifs represent the Eye of God? I have never seen this particular shape on any other box.

102

Box no. 40. Late 17th century.

103

The inspiration for the motifs of the early boxes and desks came from the Mediterranean shores, in a journey lasting for millenia. That for the later boxes came from even farther afield, but arrived here in less than fifty years.

The route was via the East India Company which in 1631 began to import hand-painted and dyed calico from the sub-continent. Hangings and bedspreads, dyed using the wax-resist process, appeared in England and became increasingly popular not only for their colour and design, but because they were relatively inexpensive. By the 1660's, a sort of Anglo-Indian style had been developed, due in part to the designs being drawn here and sent out to India for working into the finished chintz. Some of the hangings were also embroidered but together with the painted cloths these also carried the traditional motifs of vases and flowers, or flowering trees of exotic appearance.

From mid-century, these designs began to appear on the boxes, but usually in a rather blundered manner. One can understand why; whereas fine detail can easily be worked with a brush or needle, it is a different matter when it came to transferring the design onto wood, using a series of gouges. Nevertheless, there is a tremendous appeal in some of these later front panels. The designs were obviously intended to "go with" the hangings, and where these were upon beds we have an indication that the small boxes were intended for the bedroom and not the more public parts of the house.

The cheapness of the hangings made them available to a large section of the community; in 1676 there was an attempt to imitate the manufacture of them here; a patent for blockprinting "in the Indian manner" was applied for in London in that year. Such was the demand for the Indian fabrics that in 1701 the English weavers succeeded in getting further imports banned.

However, the designs were here to stay, particularly on the larger pieces. Towards the end of the century, crewel work, in which coloured wools were used, carried on the fashion, using motifs where the vegetation developed more slender stems, and semi-naturalistic trees sprang from hummocky ground, carrying large flowers and leaves; all these late designs showed strong Oriental or Chinoiserie features.

The English appetite for exotic imports was prodigious; in 1678 tapestry imports into the country alone accounted for £100,000. One can see that under this sea of patterned fabric, the box-makers of the day had no alternative but to employ motifs to match.

From about 1660 on, carving became much shallower; whether this is because of the restrictions of the design, or because it was quicker to execute is difficult to determine. In many cases this low-relief carving is not much more than an eighth of an inch deep.

Also at about this date, designs appear which repeat on each side of the lock-plate, and this "double panel" layout was to persist throughout the rest of the period in which boxes were made.

104

Box no. 41.   3rd quarter 17th century.

If it were not for the obviously country origin of this box I would have included it in the transitional period, if only because of the lunette in the central panel beneath the lock-plate. However, we are now in an entirely new territory, where the fabric designs of the bazaar have overtaken the architectural motifs of the ancient world. Here at last is the "Tree of Life" which was familiar to a number of ancient Eastern cultures, more particularly to Persia and India. The Indian form was to develop into serpentine, wandering tendrils carrying various flowers, fruits and foliage, and symbolising the life-force surging upwards from the soil. Thus in the panels on the box we can find grapes, oak and laurel leaves, and honeysuckle at the bottom corners. The oak leaves have been executed in a similar manner to those on box no. 37, (page 101), illustrating again the constraints of the shape of the gouge upon the carver. The panel also carries roses in the spandrels, and below the billowing side-roots of the trees.

The problem is to know where the carver got the pattern from. Were the oak leaves introduced in an attempt to Anglicise the design? One explanation would be that the whole thing was lifted from an imported Indian hanging which was made from a pattern sent out from this country, printed, and returned. This would probably date the box as having been made in the 1660's.

There is a further problem with Trees of Life, and that is to decide which of them we are looking at. Isaiah 11:1 says " There shall come forth a rod out of the stem of Jesse and a Branch shall grow out of his roots." In Gothic architecture this led to stone tracery in the form of the branches of a tree, The Tree of Jesse sometimes took on the form of a candleabra, and this device, or something very like it, does appear in the design of panels on some of the later boxes.

Whichever Tree one opts for, however, they all share the common habit of rising from a swelling or hummock at ground level.

Seventeenth century lining paper (or wallpaper) in the blackwork style, depicting pomegranates, carnations, and fleures-de-lis. Blackwork in embroidery was worked using a black thread upon a white linen cloth, and was popular up to the early Stuart period.

Ashmolean Museum, Oxford.

Box no.42.   Dated 1674.

Here we have a firm assurance that the box was made in 1674, the date is obviously contemporary in that it fits the design of the panel precisely and is carved in the same manner as the rest of the front. As we have seen already, dates on boxes should be regarded with suspicion, but this one is genuine.

It is also a useful benchmark for the sort of patterns which had become popular after the Restoration, the tulip in particular has finally become established as a motif.

There are many embroidery motifs on this front panel. The hummocks from which the stems arise, and those in the sub panel beneath the date, are very like the "clam" device favoured by embroiderers for decoration. The overall appearance of the panels is that of needlework, and the two-branched form of the main stems can be found on curtains and bed hangings of the period.

The pomegranates dominate the design, as well they may, being either the symbol of fertility or chastity, a useful fruit in that it capable of two opposing conceptions. The pomegranate seeds abundantly when ripe, but hides its seeds when not, the latter being favoured in Christian iconography. There is an early association of the pomegranate with the Indian Tree of Life in Italian 15th century tapestries.

Despite its association with chastity, it seems always to be depicted in an open state, revealing the seeds. Legend has it that it was the fruit that Eve gave to Adam.

In view of these Christian connections, perhaps we should allow Jesse this tree. The first owner of the box was almost certainly aware of the biblical connection.

Oddly, one rarely finds two sets of initials on the boxes, which seems to set the stamp of an single individual very firmly upon them. For the original owners they must have been very personal possessions, and one wonders why so few of the early boxes are initialled at all. To initial something is to firmly lay claim to it, so what changed over these 70 years?

Box no. 43.   Last quarter 17th century.

When I first saw this box, I wondered whether the two very striking floral patterns were Renaissance motifs, or whether there could be a connection with the Rose windows of Norman architecture, some of which were I believe called "marigold" windows.

Of course, they are none of these things; once one recognises them as lace patterns, the purpose of the box becomes clear.

The design most resembles that of "reticella" lacework, which was much in vogue towards the end of the century.

Again, beneath the lock-plate, we have a Tree, but by now it has adopted a very formalised candleabra shape, branching regularly to left and right, and unsurprisingly sprouting tulips. It still rises from a mound but the hummocks of the last example have disappeared.

The geometry of the design is very good, and the skill of the carver is obvious, but here again there is a minimum of excavation, and on closer examination, the lace roundels are simply drawn into the surface of the wood, a foretaste of things to come. Only the wood of the background of the tulip panel has been taken out, and this is very shallow.

From this time on, carving on the panels of many of the boxes degenerates into a mere outlining of the design upon the surface, with a minimum of work on the backgrounds.

Box no. 44.   Late 17th/early 18th century.

Here we come to a style where there is no attempt whatsoever to highlight the main features of the design by cutting away the background, the pattern is simply drawn upon the surface of the front panel, using a single "V" tool. The leaf forms have been slightly shaped, using a broad shallow gouge, and the veins of the leaves are also gougework. We cannot therefore excuse the carver on the grounds that he did not have the tools.

So many of these late boxes are carved in this style, which surely did not happen on grounds of economy alone, that one wonders why it was done.

The other ordinary domestic oak furniture of the eighteenth century became very plain, with a minimum of carved decoration, and perhaps these late boxes were decorated in a restrained fashion in order not to "clash" with these. Alternatively, the "in" furniture of the late 17th century employed veneers for surface effect, and deep carving must have looked very out of place in such company. To simply hint at a pattern upon the surface of the wood might have been more acceptable, particularly when one considers the intricate marquetry designs that were becoming commonplace. Very often, on larger items of furniture, these marquetry panels came in pairs, one on each side of a piece of furniture, and it cannot be coincidental that boxes made during the second half of the 17th century also employed a double-panel layout.

Box no. 45.   Late 17th/early 18th century.

Here we have a bold full-frontal design of extreme simplicity, well executed, which could not have been used during an earlier period. There is no pretence at naturalistic forms in the decoration, and there is an unfinished look to it, as if the carver suddenly decided not to bother to make even the petals more realistic, which he could have easily done by hollowing them out. This is particularly so in the demi-rose beneath the initials, which appear to have been simply sketched onto the surface of the wood.

On the other hand the work in the stand is painstakingly good, and the whole thing is a very pleasing piece of furniture. At 30" it is exactly the height of a table.

The box has a most interesting mix of timbers; the stand, the front panel, and the lid are all of oak, but everything else is of elm. All the oak is home-grown, with a maximum growth rate of 5 annual rings to the inch; the elm has a similarly high rate of growth.

One can well imagine the original owner showing it off in the hall; at 10" deep it is large enough to easily accommodate a bonnet or two.

The baluster turning of the legs of the stand is typical of the period, and helps to confirm the date of the box.

The background to the design is completely unmatted and there is no punched decoration anywhere on the box.

9 Ranunculus Batrachioides.
Frogge Crowfoote.

An illustration from Gerard's "Herbal", published in 1597. The book contains hundreds of similar drawings of English and exotic plants, all displayed as above, showing the root systems as well as the above-ground parts of the plants.

Box no. 46.   Early 18th century.

This box is very nearly the biggest in our sample, at 31" long, 17½" wide, and 10" deep. It is also a one of a group of late boxes which have a rich, deep, reddish colour, which I am sure is artificial. With furniture, there is always a reason for change, although sometimes it eludes us, probably because we do not know enough about the circumstances, social, or mechanical, at the time. With the boxes, very often competition from new fashions demanded change, and I believe the artificial colouring of wood may have been the result of the use of imported mahogany in furniture making. Although mahogany was known in Europe by the 16th century it did not really take off in England until the 18th; particularly after 1733, when tariffs against it were reduced. It was certainly around when this box was made.

The highly distinctive style of the carving, where the petals of the flowers are singled out for special treatment, occurs on a group of Welsh boxes from the Gower peninsula dated right up to the 1760's, and it is highly probable that this box is Welsh too. The rate of growth of the oak from which it was made is very high and entirely consistent with that possible in South Wales.

As to the decoration, here we have another Tree, of candleabra form, bearing the usual tulips, but at ground level the carver has included the roots.

We know that the various Herbals had been used for embroidery patterns since the 16th century, and the form of the roots in this carving could have been lifted straight out of Gerard. That the device was still being used in the 18th century is quite a tribute to the herbalist.

The whole of the background to the front panel is heavily matted, using a punch, and this gives a strong emphasis to the carving itself. If the work is Welsh, the round pellets beneath the lower branches may find a link with the traditional line and berry inlay found on some boxes made in the Vale of Glamorgan.

113

Box no. 47.

Dated 1719.

This is another "benchmark" box, in that it matches various techniques in the carving and the decoration with a firm date. I am sure that the date is contemporary with the rest of the carving, the same tool was used to carve it, and better still, the same matting punch.

It is useful to be able to put a date to the technique where only the petals of the flowers are cut into the wood, this can help us with undated boxes which use the same method of decoration. Also, on this box we find the matting on the actual surface of the wood (and not in an excavated background) which also occurs in box no. 44 on page 109. This really is an awful way to make the design stand out; on the earlier box the carver gets away with it because of the high standard of the carving of the pattern, but on this box it only serves to show up the shoddy execution of the foliage and the tulip heads.

Another point of interest in this box is the way in which the main stem rises from the bottom of the panel. This is almost exactly the same as that on box no. 45 on page 110. This sideways linking of design characteristics from dated items to others is useful, but needs caution. One is tempted to link such coincidences to a locality, but by the 18th century better communications were spreading patterns more universally. Whereas quite strong local links can be found between fixed woodwork in churches and houses and designs that appeared on furniture in the late 16th and early 17th centuries, when we get into the 18th these have become weaker.

These "petal" designs may well have a regional significance beyond Wales; perhaps Welsh carvers copied them from other parts of the Westcountry, or vice versa.

The rate of growth of the oak in all of the boxes so decorated if high, which also may indicate a Western origin.

The box was constructed with the front panel tucked in between the side panels. The end grain of these, which would otherwise have been visible from the front, has been covered by strips of half-round moulding, in the early 18th century style also found on chests of drawers.

The hasp is original to the box, which again is a very useful dating factor should this shape be found upon another box.

Such a box was obviously meant for an unsophisticated buyer, and yet, back in 1719, "M.P." must have ordered it, to have had his initials put upon the front. One wonders whether it was possible for the box maker to have bought a run of front panels with just the petals cut out, leaving the rest to be carved later by less skilled hands, and initials and dates added as required.

It would appear from this box that by 1719 the Tree of Life had been felled. However, this rustic little box is a good example of the way in which useful information can be gleaned from an unlikely source.

Box no.48.   Early 18th century.

This little box is another which has been artificially stained; it is almost exactly the colour of Cuban mahogany. It introduces a vase motif, and again, has deeply carved petals on the flowers.

I believe that this is another Welsh design; apart from the fleur-de-lis it has many features in common with the Welsh coffre-bach illustrated on page 118. The oak exhibits a very fast rate of growth, at a maximum of five annual rings to the inch, although the average is ten. The colour is very similar to that seen on other Welsh furniture of the period, which was stained with ox-blood.

The timber from which it was made is of a uniform thickness throughout, being half an inch in all the panels.

For the date, the carving is of a very high standard. The box maker however was a pessimist: the joints are dovetailed, nailed, *and* glued.

116

Box no. 49.   Dated 1736.

As one descends into the eighteenth century it is obvious that boxmaking has declined as a craft. The boxes themselves of course, are made in the same way; what has gone downhill is the carving. It is almost as if, virtually overnight, the highly skilled craftsmen who turned out the beautifully carved earlier panels had vanished without trace.

This of course is not so, they must have simply transferred their allegiance to other fields. One has to admire the carvers; they must have been a shrewd lot. Always the elite among the woodworkers, they were probably always well placed to detect and respond to changes in fashion, in that they applied to furniture that one thing which would sell it, the decoration.

It is fairly certain that no indentured carver put his hand to the front panel of the box above. And yet it has an innocent, appealing atmosphere, and no doubt when bought meant a great deal to "H.E." and "S.T." in 1736. Was it a wedding gift?

117

Box no. 50.   3rd quarter 18th century.

When it occured to the Welsh to put a drawer beneath boxes I do not know; in some of these delightful little boxes from the Gower peninsula they sometimes imitate a drawer even if there isn't one. The one illustrated above does open, the brass handles being fixed to the necks of the doves on the front of the drawer.

Vases of flowers are no doubt a universal motif, but the most likely source for them on these boxes remains the imported Indian hangings of the previous era already discussed. On this front panel they have become hybridised with the mound of earth found formerly beneath the various tree forms; the sculpted petals of the flowers and the heads of the tulip are by now old favourites.

The carver appears to have been unwilling to produce a straight line of any length, the margins of the panels sprout additional decoration in the form of semi-circles and horns, the only function of which seems to be to act as space-fillers, although it must be said that they do contribute to the overall effect of the design.

The same deep red colour already mentioned appears in the surface of the oak in this box.

The top section of the box is nine inches deep, and although traditionally the devout Welsh are held to have kept their Bibles in boxes such as these, this extreme depth makes one wonder. On the other hand, family Bibles can be immense; I have one which is six inches thick and could certainly have lived comfortably in the top compartment.

Dating these boxes can be difficult as they were made over a considerable number of years.

118

119

Box no. 51.   Late 18th century?

I have charitably placed this box at the end of the eighteenth century because of the design of the front panel, but it could possibly be Victorian. The lock-plate is the original, and is screwed to the wood, and the brass escutcheon protecting the keyhole looks machine-made.

During the Victorian "Gothic" period vast quantities of plain oak eighteenth century furniture were carved up with "Jacobean" ornamentation in a mania from which hardly anything escaped. Even the beautiful plain oak longcased clocks of earlier generations were attacked, mostly to their utter detriment.

The laurel leaves in the design of the front panel of the box may well be eighteenth century, but to me they hint of the churchyard gloom of the nineteenth. But one has to acknowledge that the carving is well done and there is a satisfying symmetry to the design. There are "C" scrolls within the tendrils which give it a slightly classical appearance.

Box no. 52.   Mid 17th century?

At first sight this looks like a rather nice plain oak desk with a drawer beneath it, which might have been made at any time during the eighteenth or nineteenth centuries. When I first saw it it was living in semi-darkness at the very back of an Antique shop, where it had stayed for so long that the owner had almost forgotten it.

Close inspection in a good light revealed the decoration on the strip of front panel above the drawer, which is virtually identical with that which appears in marginal strips upon the lids of table boxes from about 1640 to 1680. (See page 140). A further surprise was the row of punched decoration running parallel with the chevrons. The moulding section of the strip running across the panel is "ogee", which could be from either the 17th or the 18th century, but in view of the decoration, I would be quite happy with the former.

Box no. 53.   18th century?

I have modestly left this box until last as it is my own, and it has my initials on it. It is a very difficult box to date, as it has hardly any of the dating features which occur in the other fifty-two boxes in our sample.

The low relief of the carving on the panels is perhaps the only recognisable sign of a period, and about all we can say of this is that the box was probably not made before 1675; equally it could have been made at any time during the next hundred years.

The tendrils rise from the bottom of the front panel in the same way as do those in the Tree of Life examples, but instead of bearing exotic flowers and fruits they carry what I believe to be palm leaves of the coconut variety; the source of the pattern seems to have moved from the Mediterranean to the South Pacific. And what can we make of the happy little sea-monster who lives on the side panels? It has been said that this could be an Elephant Seal, which I believe is an animal from the Antarctic.

One can only speculate as to why the box was decorated in this way; perhaps the original owner was a seafarer who had been to the South Seas, as did so many during the early eighteenth century; (one brings to mind the South Sea bubble). Sailors exagerrate, and perhaps by the time he returned to relate his adventures whatever he had seen abroad had become the monster on the box. That it appears to smile is probably accidental.

Unlike all the other desks and boxes in this book, this one is made entirely from elm, and bearing in mind the appetite of woodworm for this timber, it is lucky to have survived. Perhaps it had actually travelled abroad with "T.C." and received an occasional soaking in salt water. Elm timber is resistant to sea-water, and was used in harbour construction as a consequence.

122

FOUR.

Up to this point we have been mainly concerned with decoration, and the sources which influenced it. What has emerged is that these were almost all foreign.

Until the Industrial Age, the English do not appear to have been innovators, being content, even eager to accept foreign influence upon the style of all the things which made up their immediate environment, whether in art, architecture, or fashion.

That we were to become great inventors and exploiters is another matter; perhaps we were too lazy or preoccupied to develop a decorative culture of our own. Of all the many decorative motifs on the desks and boxes, only two are "English"; the oak leaf, and the rose; even the use of the latter is suspect, as it is a universal flower.

What is English of course is the craftsmanship which went into the making of the furniture of the 16th and 17th centuries, much of it superb. This was built using the traditional skills developed during the construction of the great cathedrals and religious houses in Medieval times. One tends to think that the carpenters and woodcarvers were conservative and slow to adapt; certainly they resisted change but when this was inevitable the necessity to sell what they made was a strong incentive to produce what the new markets demanded. There can be little doubt that they learned quickly when faced with competition from foreign craftsmen.

In this last section of the book, instead of looking as the ideas behind the decoration, we look at these men who made the boxes, and the way in which they worked.

BOX MAKERS

The tools in use in the above illustration, appropriately enough from a wood-cut block, are from the previous century, but they would all have been found in a 16th century workshop. They would also all have been found in a Roman one, there had been no evolution in the shape of any of the basic hand-tools over this very considerable period of time. One can identify (from the top left) a plane, compasses, a set square, a 45 degree template, a spoke-shave, another template, a glue-pot (possibly), a frame-saw, and an axe. What may be an auger is lying on the bench. The chest that they are making is "joined".

The early box makers would probably have been astonished to learn that some of the articles which they were making in the course of their every day labour would outlast them by perhaps four hundred years, and that many would eventually be displayed in Museums. In the course of their trade they were simply fulfilling a demand and attempting to earn a living. In the cities they faced competition from other Guild members; in rural areas they were almost entirely at the mercy of the local economy.

After the Dissolution of the Monasteries their work was almost wholly secular and without the stabilising influence of the Church, "market forces" very similar to those which we have encountered in the twentieth century came into effect. When Elizabeth came to the throne, she accomplished what her father had failed to do, calling in the old debased coinage and replacing it with silver. The almost immediate effect was a stability in the price of commodities, and although prices rose throughout her reign, the increases were gradual and rents and wages were periodically adjusted. No doubt, as today, these were always below the cost of living, which was hard for the small farmer or the artisan, but for those with tangible assets a period of inflation can be no bad thing. The gradual increase in prices during the 16th century was actually beneficial in that it encouraged trade and agriculture, and thus provided work. The carpenters, joiners, turners and carvers of the late 16th and early 17th centuries must have been presented with this on an unprecedented scale.

In the cities, members of the same trade tended to congregate for convenience in the same locality, which in turn made regulation easy. The carpenters, who hitherto had been responsible for ecclesiastical structural woodwork, became house-builders and furniture makers to the nouveaux-riches, but as the demand for lighter furniture increased, the construction of which depended on the mortice-and-tenon joint, the "joyners" began to demand autonomy, and acrimonious disputes eventually led to the formation of a separate Guild for them. In 1632 the lines of demarcation were firmly fixed, and henceforth the carpenters were confined to producing furniture which was "boarded and nayled together". This is an exact description of the method of construction of the six-plank chests and boxes of the period, it also describes the work which had been the province of the Hutchier in earlier days, when their Guild had merged with that of the carpenters after the Dissolution. The success of the joiners in forming their own Guild must have been a blow to the carpenters, who were not even allowed the use of glue after 1632.

Although the construction of desks and boxes now seems obviously the work of the carpenters, the box makers became a separate branch of the Joiners' Guild in 1632, giving them a monopoly for the making of boxes and cabinets "dovetailed pinned, glued, or jointed". This does not accord with the evidence from the boxes, where until the 18th century there is hardly any use of dovetailed joints or glue, although all were "pinned".

Another anomaly lies in the apprenticeship system. A "Great Statute of Artificers" in 1563 laid down that an apprentice should serve seven years under a master craftsman, and his training should be completed by the age of twenty-four. It is difficult to see how learning to make a box should take seven years, when the rudiments could have been taught in an afternoon. As apprentices worked for bed and board and little else they were a source of virtually free labour, to the great advantage of their masters. The Guild system ensured that anyone wishing to earn a living by practising a craft had to comply, and as the Guilds had total power locally, here was no alternative if one eventually wished to make a living but to serve out an apprenticeship.

Once qualified and accepted as a member of a Guild the apprentice could either set up a workshop for himself, or become a journeyman. This latter had nothing to do with travel, although workmen could be itinerant. It simply meant a man who was paid at a daily rate, from the French "jour".

Wages were limited by statute, and were fixed by local magistrates. It is easy to see how in a locality the Guilds could become all-powerful. In theory the City Corporations controlled the Guilds but the Corporations themselves would be made up in part by members of the Guilds. As part of the same ruling class the magistrates would be unlikely to set wages at a high level, and thus the costs of production were well under control. There was little to fear from foreign competition; the Guilds in the cities had generally been successful in excluding the highly skilled French, Flemish and Dutch craftsmen until the middle of the 17th century, which probably accounts for he almost total lack of innovation in the construction and design of furniture until then.

Evelyn writing in 1660 gives us an impression of the status of the English furniture makers up to that date: "Joyners, cabinet-makers and the like, from very vulgar and pitiful artists are now come to produce works as curious for the fiting, and admirable for their dexterity in contriving, as any we meet with abroad".

By our present-day standards, the living conditions of the makers of desks and boxes must have been difficult, to say the least. From March to September the working day was from 5 am until 8 pm, with half an hour for breakfast, and an hour and a half for dinner in the afternoon. In winter they simply worked from dawn to dusk. Artificial lighting was so expensive that many working people simply went to bed at sundown.

In the cities, shops were common, but most often the maker of goods was also the retailer, living literally "over the shop". Internal communications within the country were so poor that almost all household furniture must have been made locally. One cannot imagine that heavy items such as court cupboards would have travelled far by road, when the maximum distance a coach could travel was twenty miles in a day. Similarly, the raw materials used were for the most part bought locally, although there is evidence that imported timber was available quite far inland.

Evelyn's vulgar and pitiful artists did not sign their work as the clockmakers and silversmiths of the time were required to do. Any initials found on the boxes are those of the owners. Branded or incised initials sometimes found beneath later boxes are also thought to be those of the owners, put there as part of inventories.

As to the wages paid to these anonymous craftsmen, they were hardly generous. During the construction of Wadham College between 1610 and 1613, the carpenters were paid between six and eight shillings a week. The more highly skilled carvers were paid two shillings and sixpence a day, or fifteen shillings a week. Skilled labourers earned between four and six shillings. Even the chief architect only made £1 a week, not significantly more than the carvers. Rural piecework rates by comparison were very poor; hop picking in 1614 only paid between three pence and nine pence per day.

In many respects the lot of the country maker must have been better than that of the city craftsman. The restrictions of the Guild system did not reach far into the countryside, but for that matter neither did the form of quality control which the Guilds enforced so strongly. To a certain extent he had a captive market with little competition, but his prosperity depended entirely upon that of the locality. I suspect that he also had a much more interesting job in that he would be required to make other items of furniture in addition to the little boxes. Unlike his city counterpart, he could not afford to specialise. Since Anglo-Saxon times the village carpenter had been at the heart of the rural economy, and must have remained so until well into the eighteenth century, when a better road system, and the building of canals made the swift transport of industrially-made items feasible.

By the late 16th century the prosperity of the landed classes had led to the creation of many of the great estates which were to be a feature of the English landscape for the next three hundred years, and timber was central to the construction and maintenance of these. Each large estate had its own workshops which produced not only furniture, but all the other wooden objects essential to the running of an agricultural enterprise. The larger estates must have attracted skilled carvers and woodworkers thrown out of work by the Dissolution, and many of the early desks and boxes must have been made by craftsmen who had learned their trade during the last days of the great monasteries.

Although the biggest markets for the boxes must have been in the major centres of population, a high proportion of those which remain with us today must have been made by these village and estate workmen. Towards the end of the seventeenth century this becomes more obvious as the quality of the decoration deteriorates, and the "quality" work was once again concentrated in the cities. Here, the furnishing of even the middle-class houses was beginning to sacrifice all to fashion, and the true cabinet-makers using inlay and veneer were beginning to emerge from the ranks of those who had hitherto been joiners.

LXIX.

Scriniarius

&

Tornator.

The Box-maker, and the Turner.

The

The Box-maker, 1.	*Arcularius* 1.
smootheth	
hewen-Boards 2.	edolat *Asseres* 2.
with a Plain, 3.	*Runcinâ*, 3.
upon a work board, 4.	in *Tabulâ*, 4.
he maketh them very	
smoth with	deplanat
a little plain, 5.	*Planulâ*, 5.
he boareth them tho=	perforat (terebrat)
row with an augre, 6.	*Terebrâ*, 6.
carveth them with	sculpit
a Knife, 7.	*Cultro*, 7.
fasteneth them toge=	combinat
ther with Glew,	*Glutine*
and Cramp-Irons, 8.	& *Subscudibus*, 8.
and maketh	& facit
Tables, 9.	*Tabulas*, 9.
Boards, 10.	*Mensas*, 10.
Chests 11. &c.	*Arcas* (*Ciſtas*) 11. &c.
The Turner 12. (13.	*Tornio* 12.
sitting over the treddle	sedens in *Insili*, 13.
turneth with	
a throw, 15. (14.	tornat *Torno*, 15.
upon a Turners Bench,	super *Scamno* tornato-
Bowls, 16.	Globos, 16. (rio 14.
Tops, 17.	*Conos*, 17.
Puppets, 18.	*Icunculas* 18.
and such like	& ſimilia
Turners work.	*Toreumata.*

Figulus.

Two pages from *Orbis Sensualium Pictus,* published in 1659 by Johan Amos Comenius, give us a snapshot of a boxmaker's workshop and tools at the mid-century. He has two "plains", one to "smootheth hewen boards" (note that they are not sawn), and a smaller, "to maketh then very smooth". He has an "augre" for making holes, a "Knife" for carving, and "Glew and Cramp Irons".

If one allows the artist some licence, perhaps the knife was a gouge. The only work possible with a knife is chip–carving, which is rarely found on boxes at this date.

The lower drawing shows a sort of cabinet on the right, with some open drawers, perhaps to hold nails, locks, and so on.

What is interesting is that he also makes tables and chests. A notable omission is a hammer.

CONSTRUCTION.

As has already been pointed out, there is really nothing difficult about the construction of a simple box, or its big brother the six-plank chest. All you need is six pieces of wood cut to the right length and width, a hammer, and a mouthful of nails. To prevent the wood from splitting when nailed, an auger or some sort of drill would be helpful, and for getting the surface of the boards flat a plane or a scraper would suffice. (I have seen the edge of a freshly cut piece of glass used with remarkable results).

The box makers used only three methods of joining the boards together: butting, rebating, and dovetailing; illustrated below in that order:

 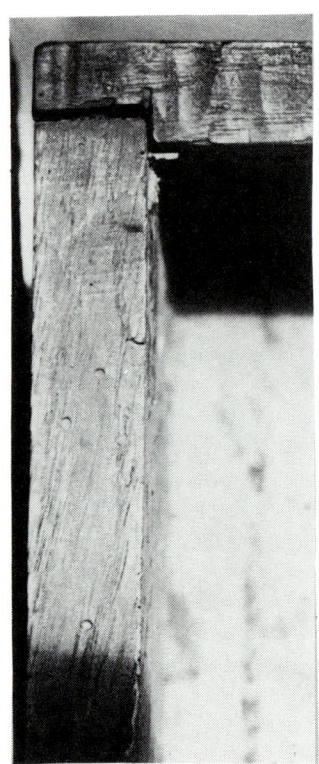 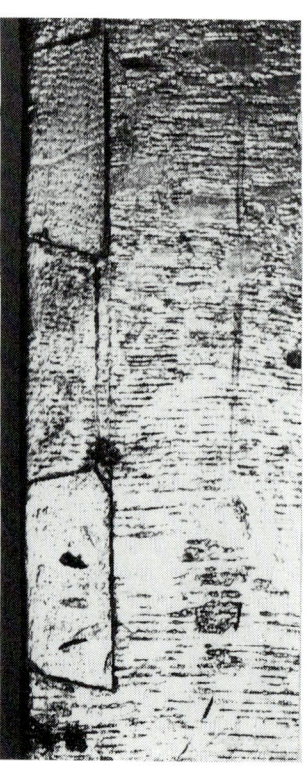

Butting the two pieces of wood together is obviously the quicker method; normally three nails completed the joint. Rebating one piece of wood into another is a much more rigid arrangement as it prevents sideways movement; the box would not attempt to become a parallelogram if one sat on it. Some early boxes and desks of quality were rebated and held together by dowels instead of nails. Most of the illustrated boxes are either butted or rebated; only one is dovetailed and this is of the eighteenth century. (Box no. 48, page 116).

One occasionally finds boxes where the front and back panels overlap those of the sides. It is difficult to understand this arrangement, as it reduces the internal dimensions of the box and really does nothing to improve the appearance. One can find Westcountry chests and boxes so constructed, but it also appears in box no 15 (page 34) the carving of which has strong connections with South Yorkshire. In this box the overlapping may have only a structural significance, as it has an odd hinge system whereby the battens beneath the lid hinge around protruberences which are actually part of the back board, and in consequence this has to protrude beyond the sides.

The long thin box no.16 (page 35) also has this feature, but contains metal hinges of the common shape.

One reason for this method of construction could well be to prevent the splitting of panels when nailed right on the edges. With very dry, well seasoned oak this is a possibility, but it could easily have been avoided by drilling out the nail holes first. Boxes with this feature never have rebated joints at the corners, the front panel is simply nailed through into the side panels.

Some later boxes have the front panels nailed between the sides, as in box no. 14 on page 33. Although this has the advantage that the nail heads are not visible from the front, the end-grain of the side panels is apparent at each side of the front. In still later boxes, this end-grain is sometimes hidden beneath half-round moulding of the sort which is found on chests of drawers of the period. A small box decorated with a line inlay in the Welsh Folk Museum at St Fagans has double half-round moulding covering the ends of the side panels. This applied moulding must have been used on the boxes to match other furniture similarly adorned and can be a useful dating factor.

In some of the boxes (generally the later ones) small tills appear. These were built into the box upon assembly, the bottoms and fronts are supported by channels cut into the back and front panels. The lids hinge upwards and in many cases support the lid of the box itself when it is open. This is also a common arrangement on the larger six-plank chests. Most of these inner lids are now missing due, I suspect to the difficulty in replacing them once their dowel hinges broke. To replace them in their original manner would have meant partially disassembling the box. I have heard these inner tills called "candle-boxes", but have never found evidence that they were thus used. No doubt their primary purpose was to separate small items such as cash and jewellery from the rest of the contents of the box.

Many of the little desks were fitted with shelves and drawers; these too were put in upon assembly. Genuine shelves were always fitted into channels in the side boards, any uprights between drawers were also rebated into the shelves and the underside of the top panels. One finds added internal fittings which are simply planted upon the panels and nailed in place.

Box no. 27, page 58.

Box no. 20, page 44.

Box no. 17, page 38.

Box no. 33, on page 95, has an interior fitted out like the inside of a Georgian bureaux, with little arcaded cubby-holes neatly arranged at the back, all these were made as one separate unit, and simply slipped into the desk, probably a hundred years after the desk itself was made.

Small drawers were fitted to many boxes, usually tucked away up under the lid; these I feel should be more accurately termed "tills" as they are too small to contain anything of any size. A typical size, from box no. 7, on page 22, is 7" long by 5½" wide, with an internal depth of one and a quarter inches. The three tills in this desk simply sit on a shelf at the back. Yew wood was a favourite for making the knobs, these are dowelled into the front panels of the tills. Many boxes still have the shelves upon which the tills sat, but very few still have their full complement, having lost them over the years. The joints in the tills are almost always butted and nailed.

One sometimes finds evidence of green timber, but always in the bottom boards. As timber seasons it shrinks slightly, always across the grain, and if an unseasoned board has been nailed the shrinking wood will draw towards the nails, causing a split between them. The narrower the board, the less shrinkage will occur, and perhaps for this reason most bottoms of desks and boxes are made from several boards.

The maximum width of board in the sample is 16", and for this to have remained stable it must have been well seasoned before use. There was no kiln drying of timber during the period of our study, oak, whether cleft or sawn would have been stacked with battens between the boards to allow air to circulate throughout the drying period. With the thinner cleft panels properly stacked and well ventilated, this could have been quite rapid. Air-drying of timber ties up capital, and there would have been considerable incentive to dry timber as quickly as possible. That craftsmen carried stocks of timber is revealed in probate valuations at the time.

Much of the timber used in the construction of desks and boxes is surprisingly thin, cleft panels in particular averaging only a quarter of an inch in many of the early examples. Could the reason for this be fiscal in that such a board would dry much more quickly than a thicker one? Until quite early in the twentieth century, cleft oak rails for gates were produced and stacked in the woods in Herefordshire, being brought into the workshops after twelve months. These rails averaged an inch in thickness, and were sufficiently seasoned when made up into gates not to show movement when eventually they went back into the open air. Given good drying conditions, a quarter inch panel might well season in six months or less.

Where several narrow boards were used to make up a lid, battens beneath them were used to join them together; these contemporary battens can usually be identified by the nails, which match those others in the box. Where lids were damaged through ill-use, one finds later battens, usually applied in a most amateur manner, glued, screwed, nailed, or in any combination of these. These later repairs are usually easily identifiable at first sight.

In some later boxes, particularly in the eighteenth century, one occasionally finds boards in lids which have been dowelled together at the edges, and then glued. This method of joining is quite sophisticated although it became a standard joinery practice. It involves the use of a jig to ensure that the dowel holes are drilled immediately opposite each other and in register in all directions. Late lids without battens beneath them made from several boards are very likely to have been joined using this method which is difficult to detect, as the dowels of course are invisible.

Very occasionally, one finds desks and boxes fitted to a stand. In the case of desks, these must have developed from the medieval armarolia, which was a lectern with an angled lid on which to put the book. Inventories of Oxford Colleges show that by the 1620's an enclosed desk in this form was fairly common, but they were not to be found among the middle classes until quite late in the 17th century, and by then they had mostly lost the stands.

Original desks and boxes fitted to contemporary stands are very rare items and "marriages" occur. Where these are well done, with the two parts of the same period, they are very difficult to detect and it is as well to view any desk attached to a stand with considerable suspicion.

In many original examples, the tops of the legs of the stand come up into the box at the corners, and the front and side panels are nailed through into them. (See box no. 9, page 26) With this arrangement, the nails from the front panel into the upright will be further from the edge of the panel than if it was nailed to the side boards in the usual way. A box which has lost its stand but was once fitted to one in this manner will have square holes in the bottom boards at the corners.

Alternatively, the box was fixed by nailing through into the tops of the rails of the stand, and here probably the only method of detecting a marriage is to look for unexplained nail holes.

This marrying of boxes and stands is probably the only faking to have taken place with them; to do it convincingly involves quite a lot of skill and cunning. That it is done at all is an indication of the value and scarcity of the originals; a skilful conjunction would involve undetectably adapting a genuine side-table of the same period which in itself would be expensive.

Honest repairs to boxes of course are another matter, and one expects to find evidence of wear and tear on wooden items which have been used for centuries. Bottom mouldings appear to have had a high casualty rate; corners too, unsurprisingly, quite often show signs of damage and repair. Apart from this, these little desks and boxes have turned out to be surprisingly robust, a tribute to simple straightforward construction in English oak.

Ring.

Butterfly.

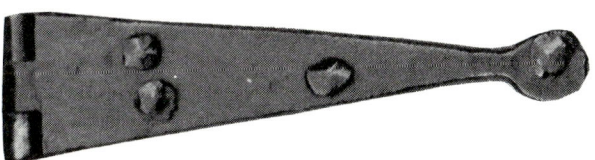

Round Fishtail.

Fishtail.

HINGES.

Hinges have no value as indicators of the age of a box; all the above were used for perhaps 200 years. The frequency of replacement however can be useful, as per the locks.

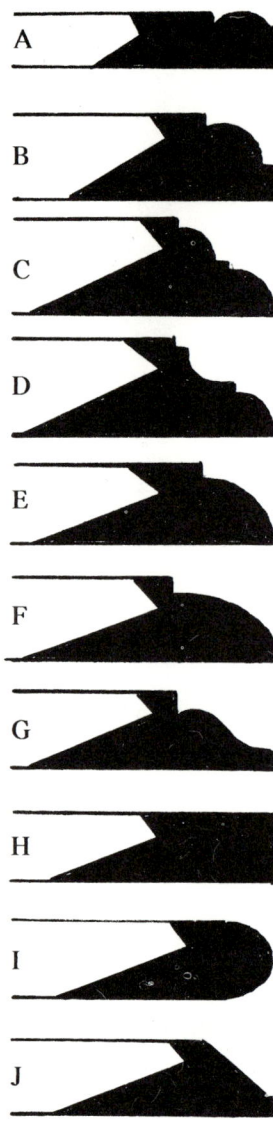

LID SURFACE/EDGE MOULDING.

(Not to scale).

136

Moulding Section.	Box numbers.
A.	9, 17, 19, 20, 34, 37.
B.	4, 5, 11, 13, 18, 21, 23.
C.	2, 35.
D.	8, 12, 24, 27, 28.
E.	1, 30, 31,42, 43.
F.	6, 14, 22, 25, 26, 33, 36, 39, 47, 49, 53.
G.	7, 10, 16, 38, 40, 46, 48, 50, 52.
H.	41.
I.	3, 15, 29, 32, 45.
J.	44, 51.

The profiles of the moulding sections employed on the lids of desks and boxes are probably one of the best indicators of age. A,B,C, and D above are all mouldings made with a scratch-block and appear on most of the early boxes. E and G are from moulding-planes, the latter appearing on many 18th century ones. F tends to be irregular, being made with a chisel. H occurs on country work, sometimes simply rounded off on the top edge only. I again, is country, but occurs throughout the period. J, a simple chamfer along the edge of the board, is generally late. The popularity of F is not surprising; this was probably done quickly by making a shallow saw cut, chiselling up to it, and rounding-off the edge.

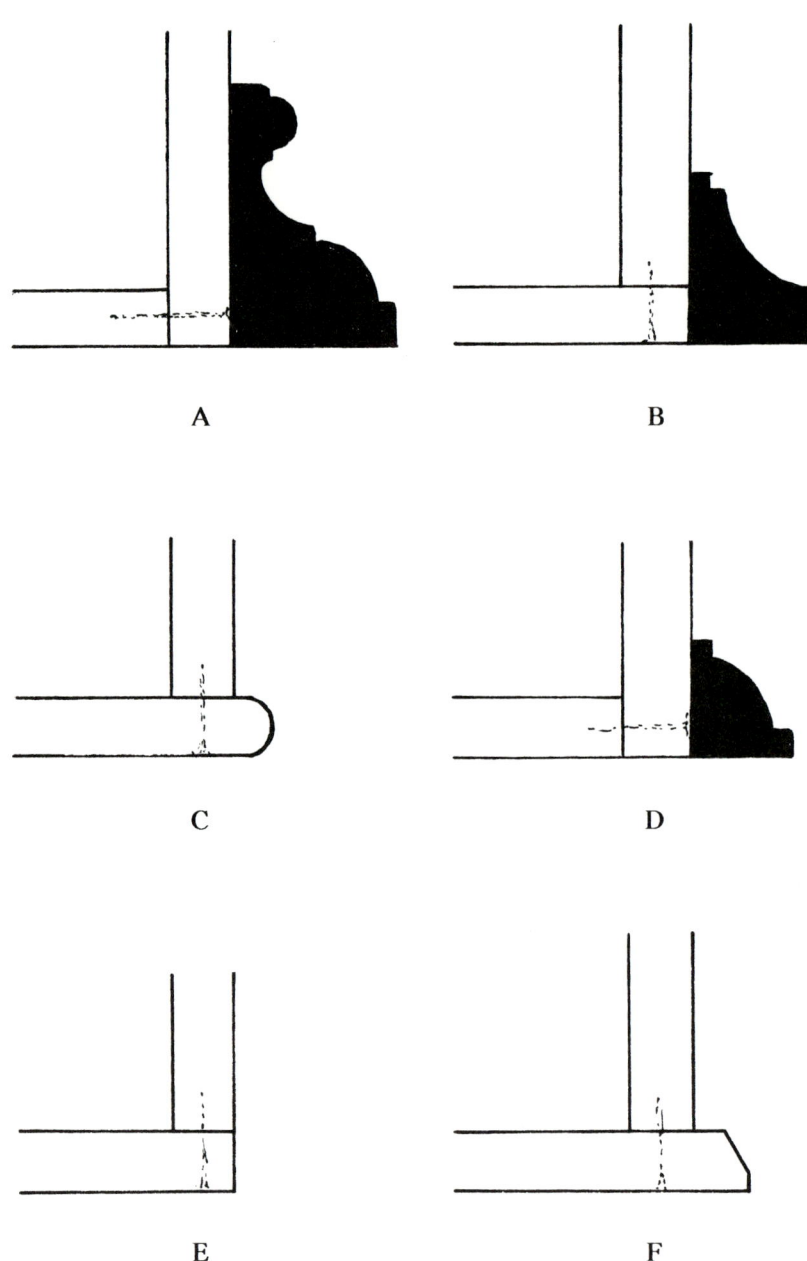

A   B

C   D

E   F

BOTTOM EDGE MOULDINGS.

(Not to scale).

Moulding section.	Box numbers.
A.	1, 18, 25, 34, 35.
B.	14.
C.	3, 9, 16, 19, 26, 31, 39, 40, 42, 46, 49, 53.
D.	4, 30, 45, 50.
E	2, 5, 7, 11, 15, 17, 20, 22, 23, 28, 32, 33, 41, 48.
F.	6, 8, 10, 12, 13, 21, 24, 27, 29, 36, 37, 38, 43, 44, 47, 51, 52.

The three shaded mouldings are all classical in origin. A has the same profile as the base moulding on architectural columns of the Ionic and Doric Orders. B is "cavetto", normally used upside down in cornices; early 18th century longcase clocks used it to hold up the hood. D is the standard "ovolo", much employed in Renaissance architecture, and in furniture-making right up to the present day.

C and F are made by extending the bottom boards and shaping them; this was a quick and easy way of finishing the bottom of a box and avoided altogether the necessity of making moulding. E occurs frequently but usually indicates that at one time the box had moulding which has fallen off. There is normally evidence that this has happened in the presence of rusted stumps of old nails at the bottom of the front and side panels.

Bottom boards were normally nailed up into the lower edges of the front, back and side panels, as in B, C, E, and F. This was a much stronger method than that in A and D, although this arrangement can occasionally be found.

Base mouldings, if original, can be a useful dating factor, but they seemed to be very vulnerable and in consequence were replaced from time to time. Replacements can be difficult to spot, as the old nail holes are covered by the new moulding, and if done a long time ago, the moulding will have picked up a patina equal to that of the rest of the box.

Scratch-block mouldings with sections similar to those used for lid decoration can be found occasionally; if original these are useful indicators of date.

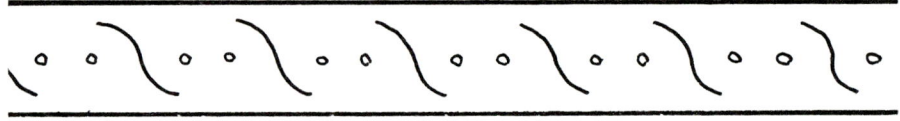

Box no. 34, page 96.

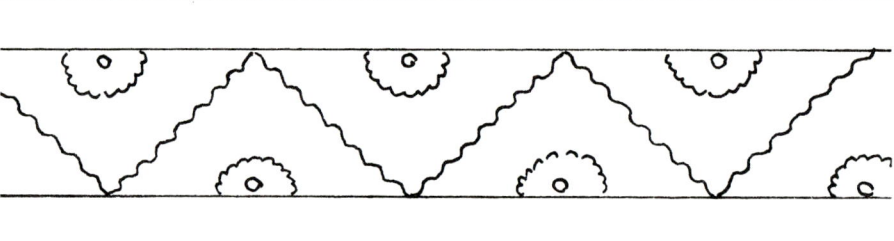

Box no. 26, page 56.

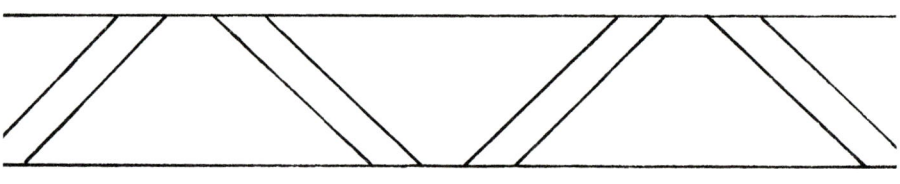

Box no. 5, page 19.

SURFACE DECORATION ON BOX LIDS.

The above are examples of the simple marginal decoration found on the lids of many boxes. The flattened "s" shapes at the top were made with the tip of a shallow gouge; the zig-zag lines in the middle example were made using a home-made punch where the design was filed into the end of a bar of soft iron. The chevrons were made with the tip of a plain chisel driven vertically into the surface of the wood. The borders of the designs were made with a scribe.

140

COSTS.

17th century inventories very often list desks and boxes, putting their value at between one and two shillings. Whether this represents the retail value or not we do not know; it may well be a secondhand value used for purposes of probate.

The value of the timber from which boxes were made is another matter; records exist showing that the price of a cubic foot of oak rose from between five and six pence in 1600 to about ten pence in 1680.

Box no. 18 (page 40), which was probably made at about the end of the 16th century, contains .57 cubic feet of planked material, and another .13 cubic feet of oak used for moulding. To the total of .7 cubic feet we should add as much as 30% for waste in conversion, and loss of sapwood; the box therefore consumed one cubic foot of round timber in the making.

As to labour costs, we have to assess just how many boxes a carpenter could make in a day. Serial production would have been a sensible method of work; i.e. instead of preparing the material for only one box, it would be economic to cut a whole run of side boards of the same width, and so on. Certainly, carvers are thought to have completed runs of carving, crosscutting these into smaller pieces as required.

Sawn oak would come into the workshop in the required thicknesses, but cleft oak would require some work to reduce the boards to a uniform thickness; cleaving results in a board with a slightly tapering cross-section.

The physical work involved in making a box includes crosscutting the ends of six boards and possibly ripping (sawing with the grain) one edge of each, assuming that one side is straight already. These eighteen operations, given skill and a well-set saw, probably took an average of three minutes each, or a total of perhaps an hour. Nailing, fitting the lock, hasp, and hinges probably took another half an hour.

Considering the length of the working day, and allowing for some work in the preparation and finishing of surfaces, a skilled carpenter could probably make five boxes a day. At a wage rate of perhaps one shilling and sixpence a day, this is about three and a halfpence per box.

The carved front panel appears to have been the most expensive item; if the carver made four panels a day (which seems reasonable; skilled carvers on repetitive work move quite quickly) the cost at two shillings and sixpence a day would be seven and a halfpence per item.

The direct costs of timber, assembly, and carving would add up to sixteen pence; to which must be added the costs of the ironwork, the workshop overheads, and the profit margin. Two shillings might well have been a reasonable retail price.

The problem of applying these costs to box no. 18 is that they really do not include the work involved in making the bottom moulding, and although no doubt this would be made as a run, such work is time consuming. With tight margins, one can well imagine that intricate mouldings would have been avoided by the makers if possible. With a box of the high quality of no. 18 there would no doubt be a premium in the purchase price over lesser boxes, but one encounters many desks and boxes where the bottom boards themselves are allowed to protrude and form the bottom moulding; see box no. 27, (page 58) where they are simply chamfered, despite the very high quality of the desk itself. This seems to indicate that margins were indeed tight.

If I am right in costing an early box at two shillings, this would indicate that the inventory prices were for secondhand items, or had been downgraded for purposes of probate valuations, a process still in vogue today.

Also, from this figure we can attempt to discover a comparable present day value.

Using the Wadham College carpenters' rate of pay at say six shillings a week, with this amount the carpenter could have afforded to buy three boxes. Today, an equally skilled man would probably earn £250 per week, making a comparable cost of a box at something over £80. Today, a brand new craftsman-made box in the old fashion, including carving would certainly cost this amount. (A good antique box would today cost perhaps £300, the added value being entirely in its attraction to a collector).

The 1600 cost of the carving is interesting, but not surprising. In a competitive world, the appearance of an item can be as important a selling-point as its function. In both the 16th and 17th centuries, possessions were symbols of status and the function of decoration was to emphasise this. The fact that we still play this game today shows how deeply embedded in human nature is the need for display. The carvers were unconsciously onto a good thing; apart from painting furniture and hanging expensive tapestry over it, how else could it become ostentatious? With a small item such as a desk or a box, carving was the only way of increasing its visual impact. It may be significant that the deepest carving is generally found on the earlier boxes; was this because only the deeper carving would show up in the poorly lit rooms of the period? As the seventeenth century progressed, rooms got lighter and smaller, and carving gradually became shallower as time went by and illumination improved.

I come back to my earlier remark, that the box makers would have been surprised to know that the things which they made so long ago, for common everyday use, would be today's collectors items. Certainly collecting was a hobby in the seventeenth century, but of things which were rare and bizarre. The idea of collecting boxes would have been extraordinary. To the makers, they were simply containers, which had to be decorated and priced to sell.

DESKS AND BOXES; THEIR ORIGINAL PURPOSE.

Desks and boxes coexisted throughout the whole period of our study, and obviously had separate uses. They were the only small lockable items of furniture available to householders, if one ignores the hanging cupboards which were mainly intended for the storage of food. The ubiquitous chest of drawers was not in general use until late in the 17th century, prior to this the only safe container for small precious items was the table box.

One assumes that initially desks were probably only owned by the wealthier members of society who could read and write, and for business reasons needed somewhere to store accounts and other papers connected with the running of large households or trading ventures. In 1601, Lady Shrewsbury had three desks at Hardwicke Hall; could this have been an early filing system, with one for "Household", one for "Estate" accounts, and perhaps the third for "Miscellaneous"? Almost all the desks illustrated originally had some arrangement of inner shelving and small drawers in order to keep separate the various things which lived in them.

In historical epics, television and the cinema have taught us that in olden days, scribes used quill pens in the same length as when they were part of a bird, but modern calligraphers find it better to cut quills to eight or nine inches long, and it seems reasonable that the earlier writers may have done the same, in which case they would have fitted easily into some of the drawers. (In passing, I read that the flight feathers of goose and turkey were the most commonly used, but those of the swan were better. These however must have been quite difficult to collect).

In the 16th century quill pens were cut "chisel" fashion, with square ends, and the "pen" knives which cut them must have lived in the boxes. I do question whether the ink was kept within them; only two of the illustrated desks showed signs of spillage. This may be evidence that the desks were carried around from place to place, which seems likely. Although one would use a desk at one's place of business, if it contained cash it would no doubt go upstairs at night; it would certainly travel with the owner if he journeyed from home.

As to writing materials, parchment and vellum in the early days were so expensive that very often they were recycled by scrubbing off the text. Also they tended to have greasy surfaces, and before being used needed to be prepared by rubbing in a "pouncing" mixture of french chalk and/or powdered resin. This too would have been kept to hand in the desks; perhaps in the smaller drawers sometimes found. As to sand to dry the ink, I am not so sure. In films one sees the nobleman scratch out the letter, dump sand on it, blow on it a few times, roll it up and pass it to the messenger saying something like: "take this to your leige poste-haste". I rather doubt the practicality of this.

The sealing of letters however was another matter, and seals and wax lived in the desks.

Writing-paper was available in England throughout the whole period of our study, the first paper-mill having been established at Hertford in 1495 by a John Tate. Paper had been made by the Chinese since the 2nd century A.D. but the more general use of it here had to wait until the printing press arrived. No doubt plain sheets of paper were stored in the desks, to keep them free of dust until needed.

Putting all this together, the desk must have been used a repository for writing implements, a filing system for accounts, a safe in which to lock away cash, and somewhere to keep correspondence. Its uses are so different from those of the boxes that it is reasonable to assume that if one was wealthy enough to need a desk, for other purposes it would have been necessary to have a box, as one could afford the possessions to put in it.

As to what these were, my thesis is that boxes belonged to the ladies in a household, whereas the men would be in charge of the desks and larger chests and coffers. Here we come again to an almost inescapable link with embroidery.

There are some very early links between carved motifs on the boxes and embroidery patterns. Alciat's "Emblems", published in Milan in 1522, and Whitney's "A Choice of Emblems" published in Leyden in the 1570's both contained patterns for wood-carvers, metal-workers, and craftsmen in general, and patterns for embroidery and lace-making were taken from them.

In the 17th century, Shorleyker's "Schole-House of the Needle" (1624), and Boler's "The Needles Excellency" were freely available in England, the latter reaching a twelfth edition by 1640. Unless they copied other embroidered items these must have been essential to the ladies of the period, and would certainly have been kept out of harms way in the boxes, together with work in progress and all the materials used.

I was shown 17th century examples of embroidery rolled up in cardboard boxes at a Museum and to my surprise realised that the size of the boxes was the same as that of the average 17th century table box.

Embroiderers must have used up pattern books fairly frequently, as the designs were pricked through onto the material from the paper, which soon disintegrated, and spare copies must have surely have been kept.

Once one accepts a link between the boxes and needlework, it is not too difficult to assume that small items of apparel would also have been kept in them. They are of a size to accept many small articles of clothing, such as gloves and handkerchiefs; these hand-made or decorated items would surely have needed to have been kept in dust-free conditions, and where else would they have been put if not into a box?

When one considers the likely possessions of the ladies of the 16th and 17th centuries there are many other things apart from clothing which needed to be kept secure, or even just out of reach of the children and servants.

Details of ladies' dress and accssories in the 1640's. From left to right: the spring fashion of 1644, English. A lady from Cologne, wearing a ruff. An English lady in the summer fashion of 1641, holding a Flemish fan. A portrait of a young girl, probably by the Dutch painter Van der Heist. An English lady dressed for the winter of 1641, wearing a muff on her left hand. A portrait of a lady, possibly French. Another English lady, in the spring fashion of 1641, with a muff in a lockable box on the table beside her. (From "Historical Encyclopaedia of Costume" by Albert Racinet. Reprinted by Bestseller Publications Ltd., Princess House, Eastcastle Street, London W1N 7AP).

What was kept in the boxes might vary according to whatever was fashionable; the late Elizabethan ladies' ruffs never attained the enormous dimensions of those used by the men and would certainly have been stored, crisply starched, in a box, together with their wire frames. A list of likely items during this period would include the expensive Spanish leather gloves fashionable up to about 1580, the small hats with their linen underbonnets, stockings of expensive foreign silk, and perhaps the bone or wooden busks which were used to stiffen the fronts of bodices.

To this one can add the valuable items of jewellery which were so important as status symbols, and all the various cosmetics so essential to appearance.

When one considers the sheer volume that these various things would occupy it is obvious that many boxes per household would be necessary if the lady were rich and fashionable, particularly as the early boxes tended to be smaller in size than the later.

As to the location of the boxes within a household, perhaps those containing items of dress, cosmetics, and so on would be found beside the bed, whereas those involved in needlework would be downstairs, wherever the lady worked by day or candlelight.

Later in the 17th century, after the advent of the "dressing" table, the bedroom boxes may well have moved downstairs into the hall, where gloves, hats, bonnets and so on could be conveniently to hand before going out.

Similarly, the lockable drawers of the chests of drawers must have usurped the function of the early boxes, and moreover solved one of the problems of storage; it was at last possible to keep things apart in one piece of furniture. A deep "bonnet drawer" was now available, and this may account for the increased depth of some of the later boxes.

Over the period of study the total number of boxes must have steadily increased as the prosperity of the country improved, and the population rose. Initially they were owned almost exclusively by the well-off, but by the end of the 17th century ownership had spread well down into the artisan classes. One forgets that boxes, being virtually indestructible, were not only bought new, but were also inherited, and no doubt were also available second-hand. However, also at the end of the century, those with fashionable pretensions would not have displayed them to public view, and when one looks at the magnificent new forms of furniture available to the better-off one can see why. Increased storage space had rendered them obsolete, and brilliant new forms of surface decoration such as marquetry had made them look dull.

One reason for their decline was possibly that they were just not big enough to accept all the accumulated possessions of the late 17th century ladies, and so they moved down the social ladder to find favour with those with less to put in them. That so many of the boxes made in the eighteenth century are "country" is surely significant.

It is sad to think that the Victorians could only find a use for them as "Bible" boxes.

SUMMARY OF DATING FACTORS.

Early.	Late.
Classical motifs: Strapwork, guilloche, lunettes, fluting. Acanthus or palm leaf infilling, acanthus or honeysuckle ends to scrolls.	Non-classical motifs based on Tree of Life. Sinuous foliage with different leaves, fruit or flowers on the same branch. Tulips appear.
Deep carving using gouges.	Shallow carving, much use of "V" tool.
Probably profuse punchwork.	Less punched decoration.
Design usually across front panel.	Designs usually two-panelled.
Boards of random thickness.	Boards sawn to equal thickness.
Some cleft oak usual.	Sawn oak usual.
Probably a coppice source.	Probably not a coppice source for oak.
Colour: honey to brown.	Colour can have a deep red hue.
Nails large and crude.	Nails small and neat.
Flat board on top of desks.	Desk lids hinged from top of backboard.
Joints usually butted or rebated.	Joints may be dovetailed or glued.
Lid moulding to early profiles page 136.	Lid moulding to late profiles page 136.

It is not until we look in detail at the work of the 16th and 17th century carvers that we can fully appreciate their skill, (or sometimes, lack of it). The modern wood-carver has at his disposal a veritable arsenal of tools; even in 1911, *Cassell's Wood Carving* listed thirty-nine basic chisels and gouges, together with a whole series of punches. From the enlarged details of the front panels from desks and boxes which follow, it would appear that if the seventeenth century carver had more than half-a-dozen he was lucky.

Moreover, they did not have the luxury of tools made from carbon-steel; what they used was probably made up by the local blacksmith from whatever passed for iron in those days.

The quality of any woodwork depends not only upon the skill of the worker, but upon the way in which the cutting tools are sharpened, and it is obvious from the following examples that they knew as much about this as we do today.

Hundreds of years of polishing of these carved surfaces has rounded-off the edges of the carving, but the assured crispness of the original work is still apparent.

(It is worthwhile looking at the side panels of boxes, if they were carved. Very often these were not so extensively polished and still retain the sharp edges left by the tool of the craftsmen who made them).

The shape of the various punches employed by the carvers is easily seen in the enlargements, particularly that of the matting-punches which produced the background to the designs.

Detail, front panel of box no. 17, page 38.

Detail, front panel of box no. 2, page 14.

Detail, front panel of box no. 39, page 102.

151

Detail, front panel of box no. 42, page 107.

Detail, front panel of box no. 31, page 64.

153

Detail, front panel of box no. 16, page 35.

154

Detail, front panel of box no. 48, page 116.

Detail, front panel of box no. 44, page 109.

GLOSSARY OF TERMS

Acanthus.	A plant indigenous to the Mediterranean; the leaf-shape of which appears as a classical motif, usually within another design.
Butted.	A joint formed by simply placing two boards at right-angles to each other and nailing them together.
Chamfer.	A bevelled surface on the edges of lids or bottom boards.
Cleft oak.	Oak logs split along the grain with a wedge or an axe to produce boards with a tapered section.
Coppice.	A forestry system whereby many crops are harvested from the same roots upon an area by regularly cutting the stems which grow from them.
Dovetail.	Joint formed by interlocking wedge-shaped sections on each of two boards.
End-grain.	The section of timber revealed by cutting across a board.
Epicormic buds.	Small shoots which appear on otherwise knot-free lengths of timber, sometimes produce "birds'-eye" appearance.
Escutcheon.	Decorative plate which protects key-holes on locks.
Fluting.	A decorative form resembling a railway viaduct with many arches. Also known as "nulling". "Stopped" fluting has an internal reed occupying the lower part.
Guilloche.	Classical motif in the form of a running ribbon, usually intertwined, with a floral centre.
Hasp.	The part of a lock which is attached to the lid of a box.
High Forest.	A "modern" forestry system whereby all the trees on an area are of the same age and height.
Honeysuckle.	Used as a classical motif; early Christian symbol of enduring faith; also associated with the Tree of Life.
Lunette.	A rainbow-shaped motif, usually with an infill of acanthus or palm leaves; sometimes with a half-rose.
Medullary ray.	Figure-marking in oak revealed by cleaving or quarter sawing.
Mortice and Tenon.	A common joint for fixing two pieces of timber to each other, the one slotting into the other.

Ogee.	A classical moulding composed in section of two curves shaped like an "S".
Ovolo.	Convex classical moulding profile, usually a full quarter of a circle in section.
Palmate.	With the shape of a palm-leaf.
Pedunculate oak.	Produces short butt lengths of timber, and crooked limbs which were used for ship and house-building.
Pomegranate.	Christian symbol of the hope of eternal life, also much used as a classical emblem in decoration.
Punchwork.	Decoration using a metal punch which transferred its shape when hammered into the surface of wood.
Quarter-sawn.	A method of cutting oak logs firstly into four segments along the grain, and then cutting boards from alternate faces from each segment, the object being to expose the radial rays within the timber.
Rebated	A joint whereby one piece of wood sits in a slot cut in another.
Reticella lace.	A form of lace-work whereby the supporting backcloth was cut away after completion.
Ropework.	Decorative feature resembling rope.
Rotation.	The period in years between felling operations.
Scratch-block.	A primitive home-made tool which produced moulding by scratching a shape into a length of timber.
Sessile oak.	Produces long lengths of clean knot-free timber; will succeed on wider range of sites than pedunculate.
Standard oak.	A tree selected by reason of straightness or vigour to grow on in a coppice area to eventually produce timber.
Strapwork.	Renaissance motif resembling flat strips of leather or iron banding, firstly used in painting to surround pictures, then architecturally, finally on furniture carving.
Volute.	Classical Greek shape derived perhaps from horns of animals, much used as architectural decoration.

ACKNOWLEDGEMENTS.

I am indebted to the many owners, past and present, of the illustrated boxes for their kind tolerance in allowing me to examine and record them, and to friends in the antique trade who have allowed me the same facility.

I would like to thank staff of the following museums for their assistance in providing illustrations for this book:

The Ashmolean Museum, Oxford.
The British Museum, London.
The Martin Von Wagner Museum, Wurzburg.
The Victoria and Albert Museum, London.
The Welsh Folk Museum, St. Fagans, Cardiff.

My particular thanks are due to Eve Finney of the Churchill Gardens Museum, Hereford, for help and advice in researching designs used also in embroidery work.

This little ink-and-wash cherub was discovered hiding behind some green
lining-paper in box no. 47, (page 114). Appropriately enough he appears to
be flourishing sprays of oak leaves. As he appeared as an end-plate in my
previous book it seems only fair to allow him space here also.

BIBLIOGRAPHY

The Styles of Ornamant: Alexander Speltz, Batsford, London 1910.

The Oxford Companion to the Decorative Arts: Ed. Harold Osborne, Oxford University Press. 1985.

Dictionary of Ornament: Philippa Lewis and Gillian Darley, Macmillan. 1986.

The Styles of European Art: Herbert Reed, Thames and Hudson 1965.

The Principles of Ornament: James Ward, Chapman and Hall, 1899.

Athenian Black Figure Vases: John Boardman, Thames and Hudson, 1974.

The Grammar of Ornament: Owen Jones, Bestseller Publications Ltd., 1987.

A Glossary of Architecture: Charles Tilt, London, 1838.

Some Main Streams and Tributaries in European Ornament from 1500 to 1750: Peter Ward-Jackson. Victoria and Albert Museum 1969.

Guide to English Embroidery: Patricia Wardle, Victoria and Albert Museum.

Costume and Fashion, a Concise History: James Laver, Thames and Hudson, 1988.

Everyday Life in Renaissance Times: E.R.Chamberlin, Batsford 1965.

Before the Industrial Revolution: Carlo M. Cipolla, Methuen 1976.

English Social History, G.M.Trevelyan, Longmans, Green and Co. Ltd., 1945

Life and Work of the People of England: Dorothy Hartley and Margaret M. Elliot: Batsford, 1925.

The Making of Britain/2: T.K.Derry and M.G. Blakeway: John Murray, London, 1969.

Manual of Traditional Woodcarving: Ed. Paul N. Hasluck: Dover Publications, New York 1977.

TREES
CHESTS
&
BOXES

OF THE SIXTEENTH AND
SEVENTEENTH CENTURIES

This unique book is the first to examine in detail
the sources of the oak from which chests, desks
and table boxes of this period were constructed.
The author reveals the wealth of evidence locked
up in the annual rings of the timber and discusses
the Forestry practices of bygone centuries which
gave rise to certain sequences of growth visible
within the timber. The study of maximum growth
rates, which are measurable in the grain of the
timber, may confirm the locality of origin of
certain styles of decoration, and from
meteorological records the author offers a method
of dating it.

Through a sequence of chests, desks and table
boxes from c. 1550 to 1720, the author illustrates
possible timber sources, methods of construction
and decoration, and offers a range of dates
between which each box was likely to have been
made. Although controversial, the book
represents a fresh approach to the problems of
assessing the date and place of manufacture of
English Oak furniture of the period.

Illustrated with the author's own photographs,
the book will appeal to all with an interest in
antique furniture, whether as dealers or
collectors, woodworkers, or indeed anyone
interested in the history of English furniture.

ISBN: 1 85421 142 0   £24.00

TREES
CHESTS
&
BOXES

OF THE SIXTEENTH AND SEVENTEENTH CENTURIES

A.J. CONYBEARE

Published by The S.P.A. Ltd.
Units 7-10
Hanley Workshops,
Hanley Road, Hanley Swan,
Worcestershire.
WR8 0DX

in conjunction with
A. J. Conybeare

Designed and produced by
Images Design and Print Ltd.

For Insurance purposes, and as an aid to identification, it is useful to record details of Antique Items. Notes should cover: overall dimensions, details of carving, locks and hasps, hinges (whether original or not), details of any repairs, original cost if known, condition, and values. Photographs or sketches should also be included.

NOTES

NOTES

NOTES

NOTES

NOTES